Springer Theses

Recognizing Outstanding Ph.D. Research

Aims and Scope

The series "Springer Theses" brings together a selection of the very best Ph.D. theses from around the world and across the physical sciences. Each volume has been selected for its scientific excellence by an internationally renowned institute or department. For greater accessibility to non-specialists, the published versions include an extended introduction, as well as a foreword by the student's supervisor explaining the special relevance of the work for the field. The series as a whole provides a valuable resource both for newcomers to the research fields, and for other scientists seeking detailed background information on special questions. Finally, it serves as an accredited documentation of the valuable contributions made by today's younger generation of scientists.

How to Apply

Theses prepared at internationally leading universities and institutes may be nominated for publication by supervisors with the endorsement of the head of department. Each such department is entitled to nominate one thesis per year. Enquiries and requests for a nomination form should be sent, together with the thesis, to Ms. Nathalie Berg (see link on right).

Eligibility Criteria

To qualify for consideration, theses should fulfil the following criteria:

- They must be written in good English.
- The topic should fall within the confines of Chemistry, Physics, Earth Sciences, Engineering and related interdisciplinary fields such as Materials, Nanoscience, Chemical Engineering, Complex Systems and Biophysics.
- The work reported in the thesis must represent a significant scientific advance.
- A maximum of 30% of the text may be a verbatim reproduction from the author's previous publications.
- They must have been examined and passed during the 12 months prior to nomination.
- The theses should have a clearly defined structure including an introduction accessible to new PhD students and scientists not expert in the relevant field.
- After acceptance, theses should be supplemented by a "Supervisor's Foreword" outlining the wider scientific relevance of the work.

Indexed by zbMATH.

Yang Geng

Data-Driven Research on Characteristics of Operating Performance of Green Office Buildings

Doctoral Thesis accepted by
Tsinghua University, Beijing, China

清華大學出版社
Tsinghua University Press

Springer

Author
Dr. Yang Geng
School of Architecture
Tsinghua University
Beijing, China

Supervisor
Prof. Borong Lin
School of Architecture
Tsinghua University
Beijing, China

ISSN 2190-5053 ISSN 2190-5061 (electronic)
Springer Theses
ISBN 978-981-95-2328-3 ISBN 978-981-95-2329-0 (eBook)
https://doi.org/10.1007/978-981-95-2329-0

Jointly published with Tsinghua University Press
The print edition is not for sale in China Mainland. Customers from China Mainland please order the print book from: Tsinghua University Press.

This Springer imprint is published by the registered company Springer Nature Singapore Pte Ltd.
The registered company address is: 152 Beach Road, #21-01/04 Gateway East, Singapore 189721, Singapore

Supervisor's Foreword

China has a large number of public buildings that have obtained green building design marks. However, people still lack a comprehensive understanding of their actual operation level, resulting in no guidance for further green building design and operation. Therefore, an in-depth understanding of the actual condition of existing green buildings has become an urgent need. The author's research on this topic has important theoretical significance and practical value for promoting the healthy development of green buildings in our country.

The author's main innovative contributions are as follows:

1. Established the largest green building operation performance database in China through the long-term monitoring of building energy consumption, indoor environmental parameters and occupant satisfaction of 63 green office buildings.
2. Revealed the macro characteristics of indoor environmental parameters of green office buildings, and explored the correlations between energy consumption, indoor environmental quality and occupant satisfaction.
3. Proposed a series of data-driven models and tools for the diagnosis and improvement of building operating performance, which were suitable for different levels of data quality.

The amount of field data and analysis work in this book is very large. The study reflects that the author has a solid basic theory and systematic in-depth specialized knowledge in the field of green building. I believe this work not only contributes to a comprehensive understanding of the actual situation of green buildings in China, but also provides an important foundation for the in-depth study in the field of green building.

Beijing, China
November 2024

Prof. Borong Lin

Supervisor's Foreword

Abstract

In recent years, the focus of green building is facing a transformation from the technical stacking at the design stage to the practical effect at the operation stage. The acquisition of actual performance data (including energy consumption, IEQ and occupant satisfaction) and its in-depth quantitative research are of great significance to comprehensively improve the quality of green buildings. However, there was a lack of high-quality, long-term and large-scale operating performance data of green buildings, which resulted in shallow and incomplete cognitions of performance characteristics and failures to guide building operation. In this book, the author carried out a series of work on data collection, characteristics cognition, performance diagnosis and optimization. An intelligent IEQ monitoring and feedback system was developed for continuous data collection with a high spatial-temporal resolution. Based on that, 63 buildings from different climate zones in China have been measured and a comprehensive green office building operating performance database was established, including energy consumption, indoor air temperature, relative humidity, CO_2, $PM_{2.5}$, illumination and occupant satisfaction. Seasonal and regional distribution characteristics of IEQ in green office buildings were revealed, as well as its quantitative correlation with energy consumption and occupant satisfaction. This work facilitated a more comprehensive and deep understanding of operating performance of green buildings. One of the remarkable discoveries was the positive and negative effects of building energy saving on indoor thermal performance, and the negative effect was found to be more significant in small office buildings with natural ventilation. Finally, a variety of data-driven models and tools with different requirements for data quality were proposed to diagnose and improve IEQ and building energy use, including: IEQ diagnosis tool via Kano model, IEQ daily pattern clustering method, building energy diagnosis model based on the regression between energy use and outdoor weather, and HVAC system smart control method based on the building load prediction model. The proposed diagnostic models and data mining methods can provide technical supports for the improvement of operating performance of green office buildings.

Keywords Green Office Buildings · Energy Consumption · Indoor Environment · Occupant Satisfaction · Building Performance Analysis · Data Mining

Acknowledgements

This study was supported by the China National Key R&D Program (Grant No. 2018YFE0106100), and the Innovative Research Groups of the National Natural Science Foundation of China (Grant No.51521005). Special thanks to my supervisor Prof. Borong Lin and other experts for their valuable advice during my doctoral research.

Declaration of the Author

This book was translated from the doctoral thesis by Dr. Yang Geng, which was accepted by Tsinghua University (China), and recognized as the 2020 outstanding Ph.D. research.

Most of the content is reorganized from the author's published journal papers, including:

[1] C. Li, Z. Lv, **Y. Geng**, H. Tang, X. Gu, B. Lin, W. Zhou, A data-driven method for the optimal control of centralized cooling station in an office park, Architectural Intelligence 2 (1) (2023) 19.
[2] **Y. Geng**, Z. Zhang, J. Yu, H. Chen, H. Zhou, B. Lin, W. Zhuang, An Intelligent IEQ Monitoring and Feedback System: Development and Applications, Engineering, 2022, 18: 218–231.
[3] **Y. Geng**, W. Ji, Y. Xie, B. Lin, W. Zhuang, A sub-sequence clustering method for identifying daily indoor environmental patterns from massive time-series data, Automation in Construction, 2022, 139: 104303.
[4] **Y. Geng**, W. Ji, Z. Wang, B. Lin, Y. Zhu, A review of operating performance in green buildings: Energy use, indoor environmental quality and occupant satisfaction, Energy and Buildings, 2019, 183: 500–514.
[5] **Y. Geng**, B. Lin, J. Yu, H. Zhou, W. Ji, H. Chen, Z. Zhang, Y. Zhu, Indoor environmental quality of green office buildings in China: Large-scale and long-term measurement, Building and Environment, 2019, 150: 266–280.
[6] **Y. Geng**, B. Lin, Y. Zhu, Comparative study on indoor environmental quality of green office buildings with different levels of energy use intensity, Building and Environment, 2019, 106482.
[7] **Y. Geng**, W. Ji, B. Lin, J. Hong, Y. Zhu, Building energy performance diagnosis using energy bills and weather data, Energy and Buildings, 2018, 172: 181–191.
[8] **Y. Geng**, J. Yu, B. Lin, Z. Wang, Y. Huang, Impact of individual IEQ factors on passengers' overall satisfaction in Chinese airport terminals, Building and Environment, 2017, 112: 241–249.

The author of this book is also the first or corresponding author of the above papers. The first two papers are open access under the terms of the Creative Commons CC

BY license, which permits unrestricted use, distribution, and reproduction in any medium. Therefore, it is not required to obtain permission to reuse these two articles. In terms of other papers, the author has got permissions (provided by Elsevier and Copyright Clearance Center) to reuse contents from these papers in this new book and the copyright license numbers are 5903941110381, 5903941371845, 5903950118298, 5903950273842, 5903950454763 and 5903950591826, respectively. In conclusion, there is no conflict of interest.

Contents

Nomenclature

ASHRAE	American Society of Heating, Refrigerating and Air-Conditioning Engineers
BREEAM	BRE Environmental Assessment Method
CASBEE	Comprehensive Assessment System for Built Environment Efficiency
CDD	Cooling Degree Day
CIBSE	Chartered Institution of Building Services Engineers
DGNB	Deutsche Gesellschaft für Nachhaltiges Bauen
DM	Data Mining
EER	Energy Efficiency Ratio
EUI	Energy Use Intensity
HDD	Heating Degree Day
HVAC	Heating, Ventilation and Air Conditioning
IAQ	Indoor Air Quality
IEQ	Indoor Environmental Quality
IMT	Inverse Modeling Toolkit
IoT	Internet of Things
ISO	International Organization for Standardization
LEED	Leadership in Energy and Environmental Design
PMP	Performance Measurement Protocols
PMV	Predicted Mean Vote
POE	Post-occupancy Evaluation
PPD	Predicted Percent Dissatisfied
USGBC	The U.S. Green Building Council

Chapter 1
Introduction

1.1 Background

Energy and health are two of the most important issues in the world today. Building acts an essential role in affecting both energy consumption and human health. On the one hand, the building industry accounts for about 40% of total energy consumption [1] and 30% of greenhouse gas emission [2]. On the other hand, people spend over 80% of their time in buildings [3] and numerous studies have verified that indoor environmental quality (IEQ) has a significant impact on occupants' comfort, health and productivity [4–6]. Therefore, the development of building industry should lead to a better IEQ for occupants with less natural resources consumption. In this context, green building has been proposed in the 1970s and rapidly developed during the last decades [7, 8]. Until now, there are more than 20 green building rating systems all over the world [9], including LEED in the United States, DGNB in Germany, BREEAM in the United Kingdom, Green Mark Scheme in Singapore, CASBEE in Japan and Green Building Evaluation Standard in China, and millions of buildings has been labeled as "green".

Along with the rapid growth of green building, an increasing number of researches have been conducted on various aspects of green building, such as assessment methods/tools, technical aspects, environmental aspects, social aspects and economic aspects [10]. At the early stage, most of the studies were design-oriented, which means that these studies mainly focused on what technologies green building "should" adopt at the design stage and what green building "should" perform during the operation. The design-oriented studies can give a guideline for the design of green building and help people know about the expected performance of green building in the future. For example, Zuo et al. [10] pointed out the crucial factors to achieve green building, i.e., technical, managerial, and behavioral categories. Li et al. [11] compared different green building assessment methods to establish a reliable and effective rating tool for green building. Wang et al. [12] developed a multi-objective genetic algorithm to identify the optimal solution for green building design. Jalaei

Y. Geng, *Data-Driven Research on Characteristics of Operating Performance of Green Office Buildings*, Springer Theses, https://doi.org/10.1007/978-981-95-2329-0_1

et al. [13] integrated building information modeling (BIM) and LEED system at the design stage of green buildings to calculate the required number of LEED certification points and to estimate the sustainability related costs.

However, the "designed" building performance usually cannot represent the actual operating result. As researches continued, more and more scholars found that there widely existed a big performance gap between the operation stage and the design stage of green building. The actual performances of many green buildings were not as good as expected, some of them performed even worse than non-green buildings. Turner et al. [14] collected the actual energy consumption data of 121 LEED certified buildings and compared them with the designed energy consumption. The results showed that 25% of green buildings failed to perform as well as designed and some green buildings even consumed more energy than the energy standard value for common buildings. Scofield [15, 16] also found that LEED-certified buildings did not save energy consumption or greenhouse gas emission compared to non-LEED buildings. Davies et al. [17] demonstrated that some green buildings in Britain did not have good indoor environment quality as expected. The reason was that these buildings generally used envelope (door or window) with high air tightness for the sake of energy-saving, however, it increased the risk of poor indoor air quality and overheated problem in transition seasons.

Due to the performance gap, design-oriented studies are not enough to reach the goal of green building. Instead, operation-oriented studies focusing on the actual operating performance of green building are attached more importance. More and more researchers begin to collect the operating performance data from the occupied green buildings and a method called post-occupancy evaluation (POE) has been widely applied. POE serves as a systematic process of evaluating buildings after they have been occupied for some time and it can identify the building's real performance, help diagnose operational problems, and finally allow the building to reach a higher level of performance [18, 19]. In POE studies, subjective surveys of occupants' satisfaction with various building aspects, such as thermal comfort, air quality, lighting, acoustics, and other non-physical factors are often conducted. Furthermore, objective environmental measurements of the occupant's space and data collection on actual energy consumption are also involved [20–22].

In recent decades, POE related studies have been conducted in many countries [23, 24], but there is still no coherent conclusion that reveals the general characteristics of operating performance of green buildings due to the lack of high-quality and large-scale measured data. Therefore, this thesis aims to give a deep and comprehensive understanding of the operating performance of green office buildings from macro to micro dimensions, based on the measured big data and a series of systematic analyses.

1.2 Literature Review

The study focuses on three aspects of operating performance of green buildings: energy use, IEQ and occupant satisfaction. The keywords used for the literature search were green building, sustainable building, operating performance, energy consumption, energy use intensity (EUI), indoor environmental quality (IEQ), occupant satisfaction, thermal comfort, indoor air quality (IAQ), measurement, questionnaire survey, post-occupancy evaluation (POE). Scopus and Web of Science, which were proved as the commonly used and credible literature searching tools [25–27], were selected to find relevant articles. As a result, a total of 925 related articles were searched, which can be used as the initial database. The author further classified the searched literatures based on the publication year, journal and countries.

(a) Year of Publication (2000–2017)

Articles were first categorized according to the year of publication to see the overall development of POE research in recent decades. The literature collected ranged from the 2000 to 2017 (Fig. 1.1). It can be clearly found that the studies of green buildings have achieved a booming development since entering the twenty-first century. The number of publications of the topic's research was 252 in 2017 and it was more than 50 times of the amount of literature in 2000, which was less than 5.

(b) Journal classification

A list of key journals with the greatest number of the topic related references are presented in Table 1.1. Several Science Citation Index (SCI) journals of high impact factors were included, such as Energy and Buildings, Renewable and Sustainable Energy Reviews, Journal of Cleaner Production, Building and Environment, Applied Energy, Energy Policy, etc.

(c) Country and area classification

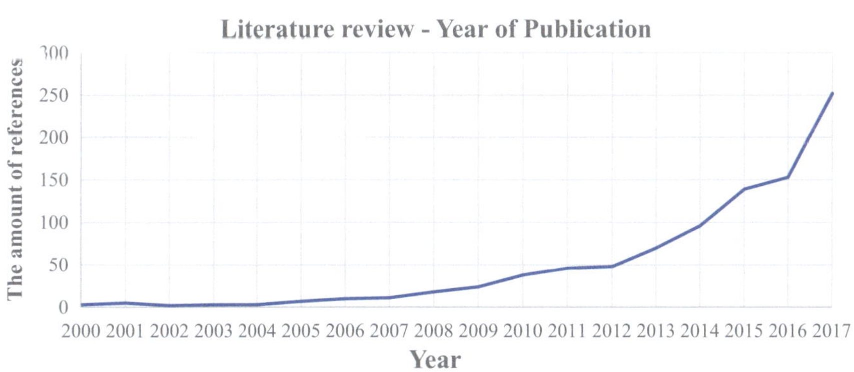

Fig. 1.1 Year of publication for the articles in the initial literature database

Table 1.1 Key journals with the greatest number of the topic related references

Journals	The number of references
Energy and Buildings	125
Renewable and Sustainable Energy Reviews	96
Journal of Cleaner Production	69
Building and Environment	57
Applied Energy	40
Sustainability	36
Journal of Green Building	34
Energy Policy	31
Building Energy Efficiency	28
Building Research and Information	25
Energy	25
Renewable Energy	22
Sustainable Cities and Society	18
Building Research Information	17
Energies	17
Indoor and Built Environment	17
Resources Science	14

Based on the authors' addresses of each paper and the function of classifying literature in Scopus and Web of Science, articles were classified by countries or regions to draw an overall picture of the geographical distribution of green building studies and literature contributions in different countries and areas (Fig. 1.2). Countries from all over the world conducted studies on green buildings and most of the literature was from Southeast Asia, Europe and North America. The United States, China, the United Kingdom, Italy and Australia were identified as the Top 5 countries with greatest contributions on green building related studies.

After classifying the initial database of articles, the author identified the most relevant 106 articles for the further review from three aspects: Measurement Protocols and Tools; Energy Consumption of Green Buildings; IEQ Performance and Occupant Satisfaction of Green Buildings. The main principles of literature screening can be summarized as follows.

- Only articles from international journals of high impact factors, research reports from authoritative institutions and articles with high citations were included.
- Only articles which have detailed operating data of building energy use, indoor environmental quality or occupant satisfaction were selected.
- Articles with just qualitative analysis or articles with only simulated data were excluded.
- Articles without strong data analyses were rejected. For example, studies without significance test were excluded for further review.

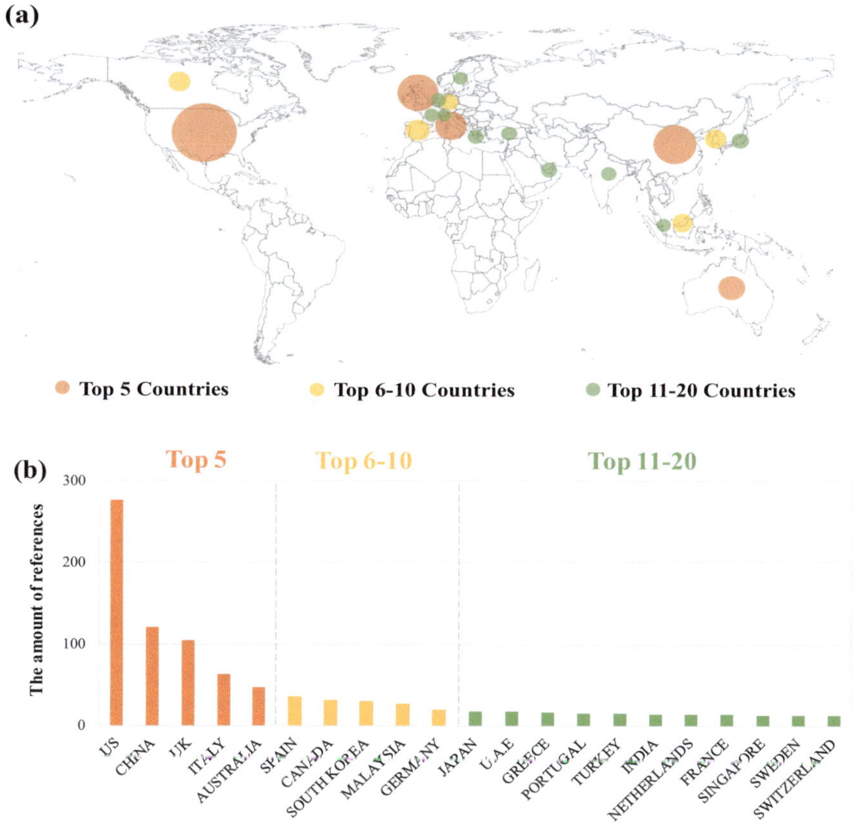

Fig. 1.2 **a** World map of the distribution of green building related references; **b** the number of references in each country (top 20)

1.2.1 Measurement Protocols and Tools

The effective measurement protocols and tools are the prerequisites for building performance data collection and further analysis. The American Society of Heating, Refrigerating and Air-Conditioning Engineers (ASHRAE)/Chartered Institute of Building Services Engineers (CIBSE)/United States Green Building Council (USGBC) have first developed Performance Measurement Protocols (PMP) for commercial buildings [28, 29]. The PMP provided a standardized and integrated protocol for measuring and comparing the overall building performance, including energy, water and IEQ. Besides that, many guidelines, protocols, and standards for the measurement of a specific building performance have also been proposed. However, without effective and efficient tools, building performance measurement was still a tough process. The followings review the data collection tools of energy, IEQ and occupant satisfaction, respectively.

- Energy data collection tools

Researchers previously used the monthly or annual utility bills [30, 31], but they were generally limited at the building level with poor quality. Recently, with the development of energy performance contracting in building sector, as well as the great development of communication and network technologies, buildings in many countries began to install smart meters and sub-meters to ensure daily or hourly energy data collection not only at the building level, but also at the system and component level [32–34]. Furthermore, the data was generally automatically recorded, stored, analyzed, and displayed on a certain platform to help energy management [35–37]. In the study of Erkoreka et al. [38], a basic monitoring system was introduced, which can measure energy consumption (including heating system and lighting system), indoor conditions and outdoor conditions. Jradi et al. [39] exhibited an online building energy monitoring tool, which can measure heating, lighting, plug and ventilation consumption continuously and can also serve as a basis of the calibrated building dynamic energy model for fault detection and diagnosis. Bonilla et al. [40] proposed a flexible and low-cost remote building energy monitoring system for employees and building managers to easily get access to the information of electrical parameters, electricity consumption and carbon footprint.

- IEQ data collection tools

The traditional IEQ monitoring tool consisted of vast independent environmental sensors [41–43], such as air temperature, humidity, CO_2, particulate matters, air pollutants, illuminance, noise, etc. (as shown in the upper part of Fig. 1.3). However, the total cost of these sensors is often overly high. Furthermore, their assembly is intensely invasive in real scenarios [44, 45], which dramatically limits the popularity of IEQ measurements on the spatial dimension. Additionally, these sensors only support local data storage, and the IEQ measurement must be conducted on the spot or in the short term [46–48]. Thus, the traditional IEQ measurement tool is deficient in the spatial–temporal scale and resolution of data.

With the rise of the internet of things (IoT) and big data technologies in recent decades, considerable progress has been made on IEQ monitoring tools [49]. New technologies, such as sensor integration, wireless data transmission, and cloud service are being introduced into the development of IEQ monitoring tools [50–52] (as shown in the lower part of Fig. 1.3). Ali et al. [53] developed an open-source platform for the integration of various IEQ sensors. This platform can support a flexible synchronization of massive and long-term IEQ measurements at a low cost. Kelly et al. [54] applied IoT to establish an IEQ monitoring system in residential buildings, which implemented the wireless interconnection of sensors via ZigBee and IPv6. De Dear et al. [55, 56] developed a continuous IEQ monitoring system, SAMBA, based on a wireless sensor network. SAMBA provided a more comprehensive solution for understanding the performance of indoor environments, which consists of both a low-cost suite of sensors and a web platform for IEQ rating and analysis. The above new tools significantly enhance the efficiency of IEQ measurement. More

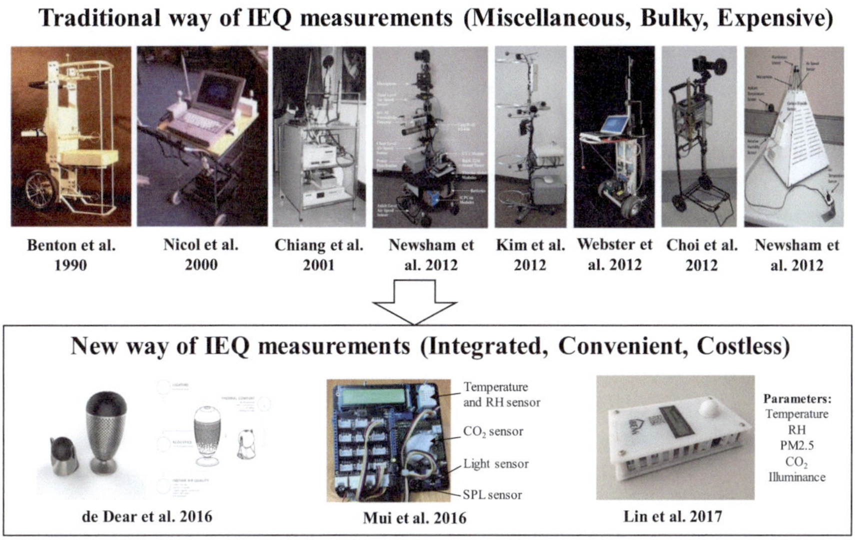

Fig. 1.3 The development of IEQ measurements, from a traditional way to a new way

importantly, they provide a possibility for the IEQ data collection with a broader spatial–temporal scale and more satisfactory resolution.

- Occupant satisfaction data collection tools

Previously, questionnaire survey on site was the most common way to collect people's satisfactions towards indoor environment [28, 57, 58]. This traditional method was a one-off solution, which meant the researchers had to go to the site again to conduct the second survey. Thus, it usually wasted lots of time and labors. The new method of acquiring subjective data can be supported by smartphones [59, 60]. Researchers can send an "invitation" containing a hyperlink to the questionnaire to the participants' smartphones and participants just need to click the screen to finish the survey. Each response is tagged with the respondent and time-stamped, which are sent back to the researchers. If the researchers want to collect the subjective data repeatedly, they only need to set a certain time point remotely and the Internet Gateway will send an alert to each participant's smartphone automatically, which reminds him/her to finish the survey.

Although the new way greatly improved the efficiency of occupant survey, the collected subjective data were hard to be synchronized with the objective monitoring IEQ parameters [61–64]. In other words, the actual IEQ condition when and where the survey occurred is unknown, and people are surveyed without understanding objective IEQ results. This defect makes further studies on the correlation between IEQ parameters and occupant satisfaction in real scenarios impossible. In sum, the existing tools cannot achieve the data collection of objective IEQ data and occupant votes simultaneously, as well as their information interaction.

1.2.2 Energy Consumption of Green Buildings

Energy consumption is one of the most concerned building performances and the reduction of building energy use is the initial driver for the green building development. Almost all green building evaluation standards provide largest credits for energy-saving design. Thus, studies on building energy performance are of great importance. Table 1.2 lists the general information and main conclusions of typical studies investigating the actual energy consumption of green buildings.

Fowler et al. [65] studied the overall sustainability of 12 General Services Administration (GSA) green buildings and collected the energy use data from utility bills for more than 12 months. The operating energy consumption was compared to the results of Commercial Buildings Energy Consumption Survey (CBECS). This study reported that the mean energy performance of the investigated green buildings was 29% better than the average value of CBECS. In the research of Baylon et al. [66], the characteristics of LEED buildings in the US Pacific Northwest was studied and compared to non-LEED buildings. The mean value of energy consumption per floor area for the 12 LEED buildings was 10% lower than the 39 non-LEED buildings of the same type and in the same region. Turner et al. [14, 67] measured energy performances of 121 LEED certified buildings in North America. The results showed that on average LEED buildings saved 25–30% energy than the national mean value for all commercial buildings. When compared to the designed energy use, the measured energy use index (EUI) presented a large amount of scatter. Over half of the buildings' actual EUI deviated by more than 25% from designed value, with 30% significantly better and 25% significantly worse. Furthermore, on average the energy saving enhanced as the certification level increased, however, the measured EUIs of different buildings within the same certification level were also dispersive. For example, the EUIs of the Gold buildings ranged from less than 63 kWh/m^2 to more than 442 kWh/m^2. Newsham et al. [68] then performed a re-analysis based on the data from the study by Turner et al. [14, 67]. The results showed that 28–35% of LEED buildings performed worse than their conventional counterparts in terms of the actual energy performance, although the average energy use per floor area of LEED buildings was 18–39% less than their conventional counterparts. Moreover, there was no clear distinction between the measured energy use of green buildings with different certification levels at design stage. Diamond et al. [69] collected the actual energy use data of 21 LEED certified buildings through utility bills. Overall, the billed energy consumption was 1% lower than simulated prediction, which was 27% compared to the baseline value. However, the actual energy performance of sample buildings had large discrepancies with the standard deviation of 46%. Furthermore, the score of LEED energy efficiency credits had no correlation with the actual energy use. The Center for Neighborhood Technology [70] studied the energy performance of 25 green buildings in Illinois. 17 of them provided complete energy data for a building, while the others only had partial energy data. The authors observed that there was a trend for lower energy consumption correlating with more energy points achieved at design stage, but there was no correlation between energy use and LEED certification

Table 1.2 The summary of typical studies investigating the actual energy consumption of green buildings

References	Year	Country	Sample size	Main conclusions
Fowler et al. [65]	2008	US	12	• The mean energy performance of the investigated green buildings was 29% better than the average value of CBECS
Baylon et al. [66]	2008	US	51 (12 Green)	• The mean value of energy consumption per floor area for the 12 LEED buildings was 10% lower than the 39 non-LEED buildings of the same type and in the same region
Turner et al. [14, 67]	2008	US	121	• On average, LEED buildings saved 25–30% energy than the national mean value for all commercial buildings • Over half of the buildings' actual EUI deviated by more than 25% from designed value, with 30% significantly better and 25% significantly worse • On average, the energy saving enhanced as the certification level increased, however, the measured EUIs within the same certification level displayed a large degree of scatter
Newsham et al. [68]	2009	Canada/US	100	• The average energy use per floor area of LEED buildings was 18–39% less than their conventional counterparts • However, 28–35% of LEED buildings performed worse than their conventional counterparts in terms of the actual energy performance • There was no clear distinction between the measured energy use of green buildings with different certification levels
Diamond et al. [69]	2006	US	21	• The actual energy consumption was 1% lower than simulated prediction • The score of LEED energy efficiency credits had no correlation with the actual energy use
Center for Neighborhood Technology [70]	2009	US	25	• There was no correlation between energy use and LEED certification level • Only 10 of 17 Whole Energy Use Projects consumed less energy than the regional average, and most buildings performed below expectations

(continued)

Table 1.2 (continued)

References	Year	Country	Sample size	Main conclusions
Li et al. [71]	2014	US/Europe/China	51	• Nearly half of the buildings did not meet the ASHRAE Standard 90.1–2004 energy target • EUIs were widely scattered in all climate zones • There was a trend toward low energy consumption in small buildings, but the correlation was not absolute
Lin et al. [72, 73]	2016	China	31	• For building type A (mix-mode), it is only in Hot-Summer Cold-Winter zone that green buildings' energy consumption is statistically significantly less than common buildings • For type B (mechanically conditioned), no statistically significantly difference between green and common has been observed in all climate zones
Scofield et al. [16]	2013	US	953 (21 Green)	• The LEED-certified buildings showed no savings when compared with non-LEED buildings • LEED Gold buildings outperformed other NYC office buildings by 20% • In contrast, LEED Silver and Certified office buildings underperformed other NYC office buildings

level. Furthermore, only 10 of 17 Whole Energy Use Projects consumed less energy than the regional average, and most buildings performed below expectations.

Now let's turn our attention to China. With the rapid development of green buildings in China, more and more Chinese scholars also attached importance to the actual energy performance of green buildings. For example, Lin et al. [72] investigated the actual energy consumption in 31 green buildings of different climate zones in China. The study indicated that for building type A (mix-mode), it was only in Hot-Summer Cold-Winter zone that the energy use of green buildings was significantly less than conventional buildings. For building type B (mechanically conditioned), there was no significant distinction of energy use between green and conventional buildings in all climate zones. Besides, the average energy use of type A and B in Hot-Summer Cold-Winter zone were found to be 15% and 23% lower than the upper limits of the Chinese standard.

The above studies on the operating energy consumption of green buildings generally reach positive conclusions, that is, green buildings have achieved certain energy saving effect overall, although the energy consumption performance of different green buildings varies greatly, and some cases have high operating energy consumption. However, some researchers held the totally negative attitude towards the energy performance of green buildings. They concluded that green buildings showed no energy savings, even more energy consumptions, as compared with non-green buildings. Among them, Prof. J. H. Scofield [15, 16, 74, 75] from Oberlin University in the United States is the representative. In terms of this disagreement on the energy performance of green buildings, the author has discussed with Prof. J. H. Scofield in *Energy and Buildings*. For more details, readers can go back to the Refs. [76, 77].

After a systematic review from different sides, the author refuted the totally negative viewpoints from several aspects [76, 77], and finally drew objective and dialectical conclusions of energy performance of green buildings:

(1) On average, the energy performance of green buildings was better than conventional buildings in the same region. However, the actual energy use of green buildings varied a lot. Some green buildings performed worse than conventional buildings, even worse than the baseline;
(2) The actual energy consumption generally had a large difference from the designed energy consumption. Many green buildings saved less energy than expected;
(3) There was no clear relationship between the actual energy use and the green building certification level.

1.2.3 IEQ Performance and Occupant Satisfaction of Green Buildings

IEQ and occupant satisfaction are also significant performance categories in the evaluation of green buildings. In general, IEQ and occupant satisfaction have a closer connection with each other. There are abundant evidences indicating that better indoor environment leads to higher occupant satisfaction, productivity, and well-being [78–81]. Furthermore, in the previous studies, objective IEQ measurements and subjective occupant satisfaction surveys were often applied together in the POE of green buildings. Therefore, in this section, studies on the IEQ and occupant satisfaction were reviewed simultaneously.

According to previous studies, the factors that affected IEQ and occupant satisfaction can be divided into physical factors and non-physical factors (Fig. 1.4). Physical factors usually consist of four parts: thermal comfort, indoor air quality, lighting, and acoustic environment, which can be evaluated by corresponding measurable parameters. Non-physical factors generally refer to those indoor qualities that are difficult to measure with instruments, including space layout, privacy, furnishing, cleanliness, facilities, and view [82]. Due to the ability of being quantified, physical factors were more frequently analyzed in the previous studies, compared to non-physical factors. Thus, the author mainly reviewed studies on physical factors in this section.

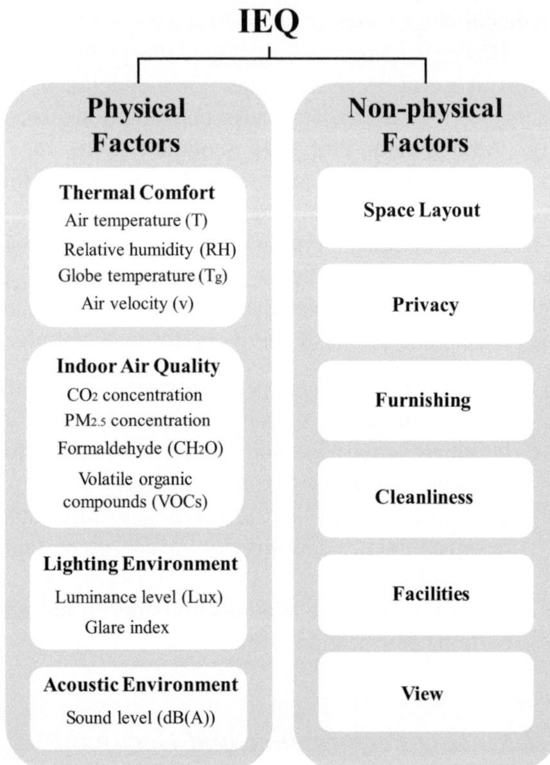

Fig. 1.4 Physical factors and non-physical factors in IEQ studies

The review found that many researchers investigated whether green buildings offered better indoor environment than conventional buildings, and in turn if occupant environmental satisfaction, productivity and health were benefitted. Most of studies combined the method of physical parameter measurement with the subjective satisfaction survey. In objective measurements, air temperature (T), relative humidity (RH), globe temperature (T_g), air velocity (v), CO_2 concentration, $PM_{2.5}$ concentration, formaldehyde (CH_2O), volatile organic compounds (VOCs), luminance level (lux) and sound level (dB(A)) are the common measured parameters. These measured data can be analyzed by theoretical model and compared to the corresponding standards. In terms of subjective survey, it is generally consisted of questions regarding to people's feeling or satisfaction with a certain IEQ factor. The occupant feedbacks are collected by the paper-based questionnaires or via online site links. There are several survey databases widely used in the world, which are listed as below:

1. BUS (Building Use Studies Occupant Survey) [83, 84].
2. COPE (Cost-effective Open-Plan Environments) [85].
3. HOPE (Health Optimization Protocol for Energy-efficient Buildings) [86].
4. CBE Survey (Centre for the Built Environment) [87].
5. BASE (Building Assessment Survey and Evaluation) [88].

6. OPN (Office Productivity Network) [89].

Table 1.3 summarizes the typical studies analyzing the IEQ performance or occupant satisfaction in green buildings in recent 10 years. Macnaughton et al. [90] investigated IEQ factors in five LEED certified buildings and five non-certified buildings in U.S. and studied the impact of IEQ on human cognitive function. The results illustrated that people in green buildings performed better and had fewer sick building symptoms than those in conventional buildings. Tham et al. [91] studied the IEQ in one Green Mark Platinum building and one non-Green Mark-certified building in Singapore. The results showed that the conventional building was about 1 °C cooler, had a lower level of illuminance, and had a 2.5–3 times higher air change rate compared to the green building. Pei et al. [62] explored the IEQ performance of green buildings in two climate zones of China through objective measurements and subjective surveys. The authors found that the objective IEQ performance of green buildings basically achieve the designed value and that green buildings were perceived more satisfied than conventional buildings. Newsham et al. [61] conducted POE studies on IEQ of 12 green and 12 non-green buildings in Canada and the northern United States. On-site physical measurements and occupant surveys indicated several environmental features which can improve occupant satisfactions, including lower background noise levels, higher light levels, conditions associated with thermal comfort, and fewer airborne particulates. Furthermore, green buildings had superior performance on human and physical factors, compared with similar non-green buildings. Deuble et al. [92] analyzed the POE survey data from two academic office buildings in Sydney, Australia and found that occupant satisfactions with IEQ were positively associated with their environmental senses, which meant that the 'green' occupants were generally better tolerated with IEQ than their 'non-green' counterparts. Kwon et al. [93] investigated the relationship between building maintenance services for IEQ and occupant satisfaction in seven office buildings in Seoul, Korea and selected two buildings for further analysis. The results showed that occupants in both offices had similar satisfactions with IEQ, although the objective IEQ performance in office E was worse than that in office B.

There were also several studies conducting large-scale surveys on occupants' satisfactions of IEQ in green buildings, without objective measurements. For instance, Schiavon and Altomonte et al. [96, 99] analyzed occupant satisfactions based on a subset of the CBE Occupant IEQ Survey database, which covered 144 buildings (65 LEED certified) with 21,477 occupant feedbacks (10,129 in LEED buildings). They found that the occupant satisfactions with IEQ in LEED and non-LEED certified buildings did not have a significant difference. Leaman et al. [106] used data from BUS occupant surveys in 177 UK buildings and explored the sources of occupant dissatisfaction with IEQ, and whether green buildings were perceived better than conventional buildings. The results indicated that the sources of dissatisfaction mainly came from temperature, air quality, lighting, and noise. Furthermore, occupants were found to be more tolerant of green buildings and the satisfaction rating for IEQ in green buildings tended to be higher than that in conventional buildings.

Table 1.3 The summary of the typical studies analyzing the overall IEQ performance in green buildings in recent 10 years

References	Year	Country	No. of buildings	Measured environmental parameters									Subjective survey samples
				T	RH	T_g	v	CO_2	$PM_{2.5}$	CH_2O or VOCs	lux	dB(A)	
Macnaughton et al. [90]	2017	US	10	√	√	–	–	√	√	√	√	√	109
Leder et al. [94]	2016	Canada	33 (12 green)	√	√	–	√	√	√	–	√	√	2545
Choi et al. [82]	2016	US	14	√	√	–	√	√	√	√	√	√	411
Candido et al. [95]	2016	Australia	18	√	√	√	√	√	–	√	√	√	2903
Tham et al. [91]	2015	Singapore	2	√	√	–	√	√	–	√	√	√	65
Pei et al. [62]	2015	China	10	√	√	–	√	√	–	–	√	√	> 1000
Schiavon et al. [96]	2014	US	144 (65 green)	–	–	–	–	–	–	–	–	–	21,477
Kemenade et al. [97]	2014	Netherlands	3	√	√	√	√	√	–	–	–	–	Not mentioned
Brown et al. [98]	2014	Canada	4	–	–	–	–	–	–	–	–	–	165
Newsham et al. [61]	2013	Canada and US	24 (12 green)	√	√	–	√	√	√	√	√	√	2545
Altomonte et al. [99]	2013	US	144 (65 green)	–	–	–	–	–	–	–	–	–	21,477
Agha-Hossein et al. [100]	2013	UK	2	–	–	–	–	–	–	–	–	–	> 300
Gou et al. [101]	2012	HK, China	2	–	–	–	–	–	–	–	–	–	99
Deuble et al. [92]	2012	Australia	2	√	-	√	–	–	–	–	–	–	155
Choi et al. [43]	2012	US	20	√	√	√	√	√	√	√	√	√	402
Kwon et al. [93]	2011	South Korea	7	√	√	√	√	√	–	–	–	–	172
Newsham et al. [102]	2009	Canada and US	1	√	√	√	√	–	–	–	√	√	89
Lee et al. [103]	2009	US	15	–	–	–	–	–	–	–	–	–	3769

(continued)

Table 1.3 (continued)

References	Year	Country	No. of buildings	Measured environmental parameters									Subjective survey samples
				T	RH	T_g	v	CO_2	$PM_{2.5}$	CH_2O or VOCs	lux	dB(A)	
Lai et al. [104]	2009	HK, China	32	√	√	√	√	√	–	–	√	√	125
Wong et al. [105]	2008	HK, China	Not mentioned	√	√	√	√	√	–	–	√	√	293
Leaman et al. [106]	2007	UK	177	–	–	–	–	–	–	–	–	–	Not mentioned

1.2.4 Summary and Comment

To sum up, the operating performance of green buildings has been investigated by many researchers and some basic findings were gained. However, there are still some issues to be solved.

- **The performance measurement tool and database**: Although the building energy monitoring system has been well-developed, there is a lack of mature tools for the data collection of both IEQ parameter and occupant satisfaction. With the rapid development of IoT technologies in recent years, some new instruments can support the multi-parameter, high-precision, high-frequency, high-density and long-term IEQ monitoring. However, these sensors were mostly at the preliminary test or case study stage without widely used in real buildings. More importantly, the existing tools cannot achieve the data collection of objective IEQ data and occupant votes simultaneously, as well as their information interaction. As a result, the comprehensive green building performance database containing a mass of energy, IEQ and occupant satisfaction data has not been established, especially in China.
- **The cognition of the operating performance characteristics of green buildings**: The basic characteristics of green buildings' energy performance and occupant satisfaction have been sufficiently revealed based on a large number of samples. However, the IEQ characteristics need a more comprehensive study. Previous IEQ studies were often limited by the short period or small-scale measurements. Such partial data cannot represent the comprehensive IEQ characteristics in green buildings due to the complex spatio-temporal patterns of IEQ. Furthermore, previous studies only focused on the single side of building performance, rather than exploring the relationship between energy, IEQ and occupant satisfaction (as shown in Fig. 1.5). This limitation makes the trade-off difficult when there is a conflict between the improving goals of different building performances.
- **The deep mining of building performance data**: With the quick accumulation of building performance data in recent years, it is more and more difficult to analyze the massive raw data based on traditional statistical or physical principle-based methods. Many previous studies just conducted some elementary analyses of building performance and missed lots of valuable information hidden behind the data. Hence, it is necessary to adopt an advanced data analysis method, namely data mining (DM). Furthermore, studies of individual building performance need to be deeper to find out problems, and to create real values that improve the operating performance.

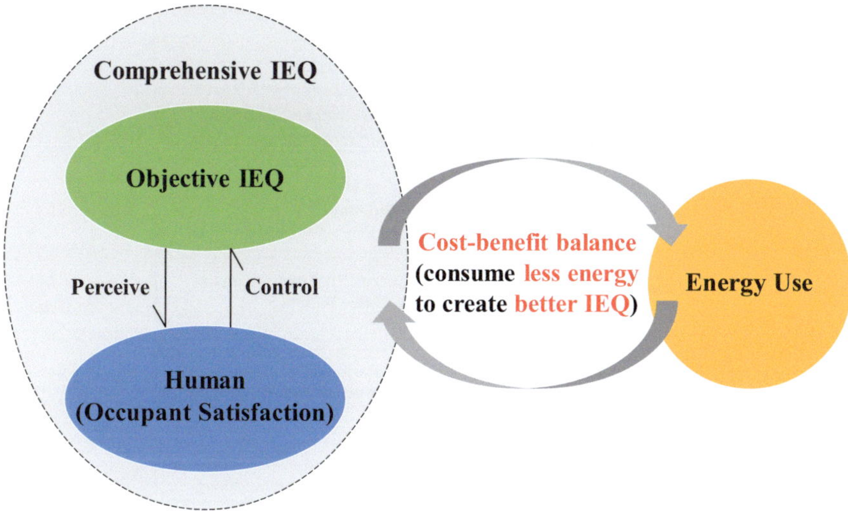

Fig. 1.5 The relationship between energy use, IEQ and occupant satisfaction

1.3 Research Contents and Framework

This thesis aims to conduct a systematic operating performance study of green office buildings from data collection to data analysis, from the characteristic cognition of regional building clusters to the deep mining of individual building. The whole thesis consists of seven chapters and they are organized as shown in Fig. 1.6.

In Chap. 2, the author developed an effective IEQ monitoring tool, which can not only achieve the large-scale and long-term continuous IEQ monitoring but also support the information interaction and occupant feedback simultaneously. Then, the author conducted a nation-wide operating performance measurement of green office buildings for years. As a result, a comprehensive database containing energy, IEQ and occupant satisfaction data of Chinese green office buildings was built up.

In Chaps. 3 and 4, the author revealed the characteristics of green office buildings from the macroscopic perspective. Chapter 3 presents the cognition of the general IEQ characteristics (single performance), while Chap. 4 illustrates the quantitative relationship between energy, IEQ and occupant satisfaction (comprehensive performance). According to Chap. 4, two questions of *how building energy conversation affects IEQ performance* and *how objective IEQ performance affects occupant satisfaction* can be answered.

In Chaps. 5 and 6, the author mainly focused on the performance of individual building and used the data to improve building operation level (microscopic perspective). The two chapters aim at different quality of data, respectively. Chapter 5 provides an elementary diagnosis solution based on poor-quality data, while Chap. 6 describes how to fully exploit the high-quality operating performance data through DM.

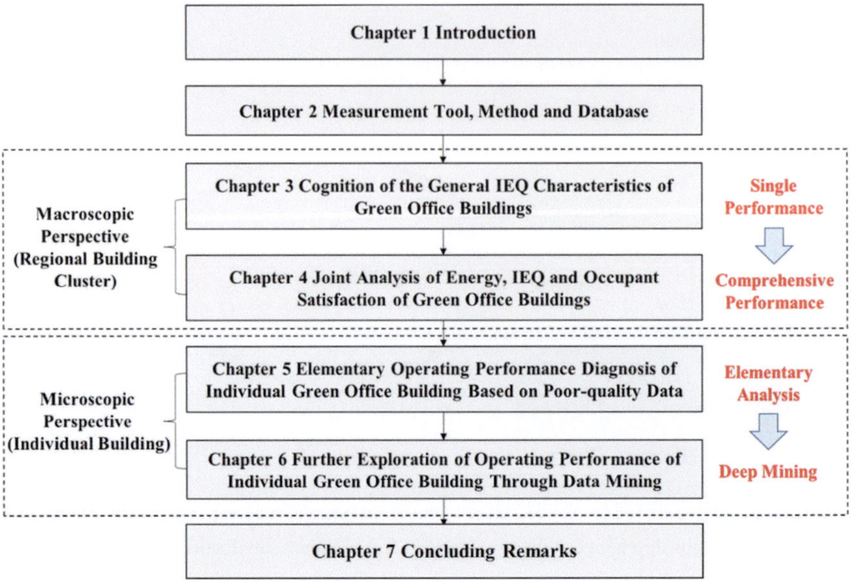

Fig. 1.6 Thesis framework

In Chap. 7, the key achievements and findings of the whole thesis were summarized.

References

1. WBCSD (2008) Energy efficiency in buildings, business realities and opportunities. The world business council for sustainable development. http://sustainca.org/sites/default/files/EEffPu-WBCSD.pdf
2. UNEP SBCI, Buildings and Climate Change, United Nations Environment Programme, 2009
3. GeoMet (1989) Report to Congress on indoor air quality. Volume 2: Assessment and control of indoor air pollution, US Environmental Protection Agency; Office of Air and Radiation, Washington, DC
4. Spengler JD, Sexton K (1983) Indoor air pollution: a public health perspective. Science 221(4605):9–17
5. Mendell MJ, Fisk WJ, Kreiss K, Levin H, Alexander D, Cain WS, Girman JR, Hines CJ, Jensen PA, Milton DK, Rexroat LP, Wallingford KM (2002) Improving the health of workers in indoor environments: priority research needs for a National Occupational Research Agenda. Am J Public Health 92:1430–1440
6. Weschler CJ (2009) Changes in indoor pollutants since the 1950s. Atmos Environ 43:153–169
7. Holmes MJ, Hacker JN (2007) Climate change, thermal comfort and energy: meeting the design challenges of the 21st century. Energy Build 39:802–814
8. Amos D, Chan APC (2016) Critical analysis of green building research trend in construction. J Habitat Int 57:53–63. https://doi.org/10.1016/j.habitatint.2016.07.001
9. Zhang Y, Wang J, Hu F, Wang Y (2017) Comparison of evaluation standards for green building in China, Britain, United States. Renew Sust Energ Rev 68:262–271

10. Jian Z, Yu ZZ (2014) Green building research-current status and future agenda: a review. Renew Sustain Energy Rev 30:271–281. https://doi.org/10.1016/j.rser.2013.10.021
11. Li Y, Chen X, Wang X, Xu Y, Chen P-H (2017) A review of studies on green building assessment methods by comparative analysis. Energy Build 146:152–159
12. Wang W, Zmeureanu R, Rivard H (2005) Applying multi-objective genetic algorithms in green building design optimization. Build Environ 40:1512–1525
13. Jalaei F, Jrade A (2015) Integrating building information modeling (BIM) and LEED system at the conceptual design stage of sustainable buildings. Sustain Cities Soc 18:95–107
14. Turner C (2008) Energy performance of LEED® for new construction buildings. New Buildings Institute Green Buildings Council
15. John H (2009) Scofield, Do LEED-certified buildings save energy? Not really. Energy Build 41:1386–1390
16. John H (2013) Scofield, Efficacy of LEED-certification in reducing energy consumption and greenhouse gas emission for large New York City office buildings. Energy Build 67:517–524
17. Davies M, Oreszczyn T (2012) The unintended consequences of decarbonising the built environment: a UK case study. Energy Build 46:80–85
18. Preiser WFE, Rabinowitz HZ, White ET (1998) Post occupancy evaluation. Van Nostrand Reinhold, New York
19. Birt B, Newsham GR (2009) Post-occupancy evaluation of energy and indoor environment quality in green buildings: a review
20. Cohen R, Standeven M, Bordass B, Leaman A (2001) Assessing building performance in use 1: the probe process. Build Res Inf 29:85–102
21. Bordass B, Cohen R, Standeven M, Leaman A (2001) Assessing building performance in use 2: technical performance of the probe buildings. Build Res Inf 29:103–113
22. Fergus N, Susan R (2005) Post-occupancy evaluation and field studies of thermal comfort. Build Res Inf 33(4):338–346
23. Meir IA, Garb Y, Jiao D, Cicelsky A (2009) Post-occupancy evaluation: an inevitable step toward sustainability. Adv Build Energy Res 3:189–219
24. Nawawi AH, Khalil N (2008) Post-occupancy evaluation correlated with building occupants' satisfaction: an approach to performance evaluation of government and public buildings. J Build Apprais 4:59–69
25. Mckenzie D (1998) Web of Science. Bull Med Libr Assoc 86:629
26. Bakkalbasi N, Bauer K, Glover J, Wang L (2006) Three options for citation tracking: Google Scholar, Scopus and Wen of Science. Biomed Digit Lib 3:1–8
27. Falagas ME, Pitsouni EI, Malietzis GA, Pappas G (2008) Comparison of PubMed, Scopus, Web of Science, and Google Scholar: strengths and weakness. FASEB J 22:338–342
28. ASHRAE/CIBSE/USGBC (2010) Performance measurement protocols for commercial buildings. American Society of Heating, Refrigeration, and Air Conditioning. Engineers, Atlanta
29. ASHRAE (2012) Performance measurement protocols: best practices guide. American Society of Heating, Refrigerating and Air-Conditioning Engineers, Atlanta
30. Dong B, Lee SE, Sapar MH (2005) A holistic utility bill analysis method for baselining whole commercial building energy consumption in Singapore. Energy Build 37:167–174
31. ASHRAE, Measurement of Energy and Demand Savings (2002) ASHRAE Guideline 14
32. Ahmad MW, Mourshed M, Mundow D, Sisinni M, Rezgui Y (2016) Building energy metering and environmental monitoring—a state-of-the-art review and directions for future research. Energy & Buildings 120:85–102
33. Wei N, Yong W, Yan S et al (2009) Government management and implementation of national real-time energy monitoring system for China large-scale public building. Energy Policy 37:2087–2091
34. Zhao L, Zhang JL, Liang RB (2013) Development of an energy monitoring system for large public buildings. Energy Build 66:41–48
35. Molina-Solana M, Ros M, Ruiz MD, Gómez-Romero J, Martin-Bautista MJ (2017) Data science for building energy management: a review. Renew Sustain Energy Rev 70:598–609

36. Scarpa M, Ravagnin R, Schibuola L, Tambani C (2017) Development and testing of a platform aimed at pervasive monitoring of indoor environment and building energy. Energy Procedia 126:282–288
37. Papantoniou S, Mangili S, Mangialenti I (2017) Using intelligent building energy management system for the integration of several systems to one overall monitoring and management system. Energy Procedia 111:639–647
38. Erkoreka A, Garcia E, Martin K, Teres-Zubiaga J, Portillo DL (2016) In-use office building energy characterization through basic monitoring and modelling. Energy Build 119:256–266
39. Jradi M, Arendt K, Sangogboye FC, Mattera CG, Markoska E, Kjærgaard MB, Veje CT, Jørgensen BN (2018) ObepME: an online building energy performance monitoring and evaluation tool to reduce energy performance gaps. Energy Build 166:196–209
40. Bonilla D, Samaniego MG, Ramos R, Campbell H (2018) Practical and low-cost monitoring tool for building energy management systems using virtual instrumentation. Sustain Cities Soc 39
41. Chiang CM, Chou PC, Lai CM, Li YY (2001) A methodology to assess the indoor environment in care centers for senior citizens. Build Environ 36:561–568
42. Kim H, Haberl JS (2012) Field-test of the new ASHRAE/CIBSE/USGBC performance measurement protocols for commercial buildings: basic level. ASHRAE Trans:135–142
43. Choi JH, Loftness V, Aziz A (2012) Post-occupancy evaluation of 20 office buildings as basis for future IEQ standards and guidelines. Energy Build 46:167–175
44. Heinzerling D, Schiavon S, Webster T, Arens E (2013) Indoor environmental quality assessment models: a literature review and a proposed weighting and classification scheme. Build Environ 70:210–222
45. Geng Y, Ji W, Wang Z, Lin B, Zhu Y (2019) A review of operating performance in green buildings: energy use, indoor environmental quality and occupant satisfaction. Energy Build 183:500–514
46. Leder S, Newsham GR, Veitch JA, Mancini S, Charles KE (2016) Effects of office environment on employee satisfaction: a new analysis. Build Res Inf 44:34–50
47. Tang H, Ding J, Li C, Li J (2019) A field study on indoor environment quality of Chinese inpatient buildings in a hot and humid region. Build Environ 151:156–167
48. Zuhaib S, Manton R, Griffin C, Hajdukiewicz M, Keane MM, Goggins J (2018) An Indoor Environmental Quality (IEQ) assessment of a partially-retrofitted university building. Build Environ 139:69–85
49. Coulby G, Clear A, Jones O, Godfrey A (2020) A scoping review of technological approaches to environmental monitoring. Int J Environ Res Public Health 17
50. Martín-Garín A, Millán-García JA, Baïri A, Millán-Medel J, Sala-Lizarraga JM (2018) Environmental monitoring system based on an Open Source Platform and the Internet of Things for a building energy retrofit. Autom Constr 87:201–214
51. Karami M, McMorrow GV, Wang L (2018) Continuous monitoring of indoor environmental quality using an Arduino-based data acquisition system. J Build Eng 19:412–419
52. Perez AO, Bierer B, Scholz L, Wöllenstein J, Palzer S (2018) A wireless gas sensor network to monitor indoor environmental quality in schools. Sensors (Switzerland) 18
53. Ali AS, Zanzinger Z, Debose D, Stephens B (2016) Open Source Building Science Sensors (OSBSS): a low-cost Arduino-based platform for long-term indoor environmental data collection. Build Environ 100:114–126
54. Kelly SDT, Suryadevara NK, Mukhopadhyay SC (2013) Towards the implementation of IoT for environmental condition monitoring in homes. IEEE Sens J 13:3846–3853
55. Parkinson T, Parkinson A, de Dear R (2019) Continuous IEQ monitoring system: context and development. Build Environ 149:15–25
56. Parkinson T, Parkinson A, de Dear R (2019) Continuous IEQ monitoring system: performance specifications and thermal comfort classification. Build Environ 149:241–252
57. Goto T, Mitamura T, Yoshino H, Tamura A, Inomata E (2007) Long-term field survey on thermal adaptation in office buildings in Japan. Build Environ 42:3944–3954

58. Wagner A, Gossauer E, Moosmann C, Gropp T, Leonhart R (2007) Thermal comfort and workplace occupant satisfaction—results of field studies in German low energy office buildings. Energy Build 39:758–769
59. de Dear R, Kim J, Parkinson T (2018) Residential adaptive comfort in a humid subtropical climate—Sydney Australia. Energy Build 158:1296–1305
60. Nian FZ, Wang K, Zhao QC (2017) Research on indoor environmental comfort based on complaints. Int J Innov Comp Inf Control 13:1323–1333
61. Newsham GR, Birt BJ, Arsenault C, Thompson AJL, Veitch JA, Mancini S, Galasiu AD, Gover BN, Macdonald IA, Burns GJ (2013) Do 'green' buildings have better indoor environments? New evidence. Build Res Inf 41:415–434
62. Pei Z, Lin B, Liu Y, Zhu Y (2015) Comparative study on the indoor environment quality of green office buildings in China with a long-term field measurement and investigation. Build Environ 84:80–88
63. Lee JY, Wargocki P, Chan YH, Chen L, Tham KW (2019) Indoor environmental quality, occupant satisfaction, and acute building-related health symptoms in Green Mark-certified compared with non-certified office buildings. Indoor Air 29:112–129
64. Oldham E, Kim H (2020) IEQ field investigation in high-performance, urban elementary schools. Atmosphere 11
65. Fowler K, Rauch E (2008) Assessing green building performance: a post occupancy evaluation of 12 GSA buildings
66. Baylon D, Storm P (2008) Comparison of commercial LEED buildings and non-LEED buildings within the 2002–2004 Pacific Northwest commercial building stock
67. Turner C (2006) LEED building performance in the Cascadia Region: a post occupancy evaluation report
68. Newsham GR, Mancini S, Birt BJ (2009) Do LEED-certified buildings save energy? Yes but …. Energy Build 41:897–905
69. Diamond R, Opitz M, Hicks T, Von Neida B, Herrera S (2006) Evaluating the energy performance of the first generation of LEED-certified commercial buildings. Office of Scientific & Technical Information Technical Reports
70. Center for Neighborhood Technology (2009) Regional green building case study project: a post occupancy study of LEED projects in Illinois
71. Li C, Hong T, Yan D (2014) An insight into actual energy use and its drivers in high-performance buildings. Appl Energy 131:394–410
72. Lin BR, Liu YC, Wang Z, Pei ZF, Davies M (2016) Measured energy use and indoor environment quality in green office buildings in China. Energy Build 129:9–18
73. Liu YC (2018) Research on the measurement and evaluation of energy consumption and indoor environment quality of green office buildings. Ph.D. thesis. Tsinghua University, Beijing
74. Scofield JH (2009) A re-examination of the NBI LEED building energy consumption study. In: Proceedings of the international energy program evaluation conference (IEPEC) (Portland, OR, August 12–14, 2009). http://www2.oberlin.edu/physics/Scofield/pdf_files/Scofield%20IEPEC%20paper.pdf
75. Scofield JH, Doanes J (2018) Energy performance of LEED-certified buildings from 2015 Chicago benchmarking data. Energy Build 174(Sept 2018):402–413. https://doi.org/10.1016/j.enbuild.2018.06.019
76. Geng Y, Ji W, Wang Z, Lin B, Zhu Y (2019) Response to the commentary on "A review of operating performance in green buildings: Energy use, indoor environmental quality and occupant satisfaction" by John H. Scofield. Energy Build 194:366–368
77. Scofield JH (2019) Comment on the paper entitled, "A review of operating performance in green buildings: Energy use, indoor environmental quality and occupant satisfaction" by Geng et al. Energy Build 194:369–371
78. Newsham GR, Veitch JA, Charles KE (2008) Risk factors for dissatisfaction with the indoor environment in open-plan offices: an analysis of COPE field study data. Indoor Air 18:271–282
79. Thayer JF, Verkuil B, Brosschot JF (2010) Effects of the physical work environment on physiological measures of stress. Eur J Cardiovasc Prev Rehabil 17:431–439

80. Al Horr Y, Arif M, Kaushik A, Mazroei A, Katafygiotou M, Elsarrag E (2016) Occupant productivity and office indoor environment quality: a review of the literature. Build Environ 105(2016):369–389
81. Fisk WJ (2000) Health and productivity gains from better indoor environments and their relationship with building energy efficiency. Annu Rev Energy Env 25:537–566
82. Choi JH, Moon J (2016) Impacts of human and spatial factors on user satisfaction in office environments. Build Environ 114:23–35
83. Dykes C, Baird G (2013) A review of questionnaire-based methods used for assessing and benchmarking indoor environmental quality. Intell Build Int 5:135–149
84. Leaman A, Bordass B (2001) Assessing building performance in use 4: the Probe occupant surveys and their implications. Build Res Inf 29:129–143
85. Newsham G (2005) Cost-effective Open-Plan Environments (COPE) project
86. Bluyssen PM, Aries M, van Dommelen P (2011) Comfort of workers in office buildings: the European HOPE project. Build Environ 46:280–288
87. Zagreus L, Huizenga C, Arens E, Lehrer D (2004) Listening to the occupants: a Web-based indoor environmental quality survey. Indoor Air 14:65–74
88. Levermore GJ, Lowe DJ, Ure JW (1999) Occupant feedback questionnaire producing a fingerprint and a score. ASHRAE Trans 105:661
89. Oseland N (2004) Occupant feedback tools of the office productivity network. http://www.officeproductivity.co.uk
90. Macnaughton P, Satish U, Laurent JGC, Flanigan S, Vallarino J, Coull B, Spengler JD, Allen JG (2017) The impact of working in a green certified building on cognitive function and health. Build Environ 114:178–186
91. Tham KW, Wargocki P, Tan YF (2015) Indoor environmental quality, occupant perception, prevalence of sick building syndrome symptoms, and sick leave in a Green Mark Platinum-rated versus a non-Green Mark-rated building: a case study. Sci Technol Built Environ 21:35–44
92. Deuble MP, de Dear RJ (2012) Green occupants for green buildings: the missing link? Build Environ 56:21–27
93. Kwon SH, Chun C, Kwak RY (2011) Relationship between quality of building maintenance management services for indoor environmental quality and occupant satisfaction. Build Environ 46:2179–2185
94. Leder S, Newsham GR, Veitch JA, Mancini S, Charles KE (2016) Effects of office environment on employee satisfaction: a new analysis. Build Res Informat 44:34–50
95. Candido C, Kim J, de Dear R, Thomas L (2016) BOSSA: a multidimensional post-occupancy evaluation tool. Build Res Informat 44:214–228
96. Schiavon S, Altomonte S (2014) Influence of factors unrelated to environmental quality on occupant satisfaction in LEED and non-LEED certified buildings. Build Environ 77:148–159
97. Kemenade PV, Loomans MGLC, Opschoor S, Hensen JLM (2014) Building comfort performance assessment using a monitoring tool
98. Brown C, Gorgolewski M (2014) Assessing occupant satisfaction and energy behaviours in Toronto's LEED gold high-rise residential buildings. Int J Energy Sect Manage 8:492–505
99. Altomonte S, Schiavon S (2013) Occupant satisfaction in LEED and non-LEED certified buildings. Build Environ 68:66–76
100. Agha-Hossein MM, El-Jouzi S, Elmualim AA, Ellis J, Williams M (2013) Post-occupancy studies of an office environment: energy performance and occupants' satisfaction. Build Environ 69:121–130
101. Gou ZH, Lau SSY, Shen J (2012) Indoor environmental satisfaction in two LEED offices and its implications in green interior design. Indoor Built Environ 21:503–514
102. Newsham G, Brand J, Donnelly C, Veitch J, Aries M, Charles K (2009) Linking indoor environment conditions to job satisfaction: a field study. Build Res Informat 37:129–147
103. Lee YS, Guerin DA (2009) Indoor environmental quality related to occupant satisfaction and performance in LEED-certified buildings. Indoor Built Environ 18:293–300

104. Lai ACK, Mui KW, Wong LT, Law LY (2009) An evaluation model for indoor environmental quality (IEQ) acceptance in residential buildings. Energy Build 41:930–936
105. Wong LT, Mui KW, Hui PS (2008) A multivariate-logistic model for acceptance of indoor environmental quality (IEQ) in offices. Build Environ 43:1–6
106. Leaman A, Bordass B (2007) Are users more tolerant of 'green' buildings? Build Res Inf 35:662–673

Chapter 2
Measurement Tool, Method and Database

2.1 Background

The first step to carry out quantitative research on the operation performance of green buildings is the basic data collection and database establishment, including energy consumption, IEQ and occupant satisfaction. Western developed countries have earlier built building energy and IEQ databases and accumulated large-scale data, such as CBECS in the United States, BPIE in the European Union and DECC in Japan (building energy databases); CBE in the United States, COPE in Canada and BOSSA in Australia (IEQ databases). In recent years, some countries have also begun to gradually integrate subjective and objective data of building performances, such as NEAT in the United States and BUS-PROBE in the United Kingdom. However, due to the limitations of traditional monitoring tools and methods, indoor environmental parameters in the above databases often have large shortcomings in terms of temporal and spatial distribution. In China, the post-occupancy evaluation of building performances started late. There is a lack of systematic practice experience and large-scale data accumulation, resulting in the fact that no high-quality green building operating performance database has been established yet. Under this background, it is impossible to carry out a systematic and in-depth quantitative research.

In order to solve the above problem, the author developed a new building performance monitoring tool. Based on this tool, long-term building performance data of many green office buildings were collected nationwide, a comprehensive building performance database was established. This work fills the lack of basic data on the actual performance of green office buildings in China, and lays a solid data foundation for the following data-driven study on building performances.

Y. Geng, *Data-Driven Research on Characteristics of Operating Performance of Green Office Buildings*, Springer Theses, https://doi.org/10.1007/978-981-95-2329-0_2

2.2 Development of the IBEM

The author has developed the Intelligent Built Environment Monitor (IBEM) since 2014. Up to now, IBEM is still in continuous improvement and upgrading. This section introduces the whole architecture of IBEM, including both hardware and software.

As shown in Fig. 2.1, IBEM is built on five parts: Integrated sensor, Cloud server, Database, Web platform, and Mobile interface, which are connected and finally evolve as a pathway of dataflow. IEQ data (including air temperature, relative humidity, CO_2, $PM_{2.5}$, and illuminance) are collected by the integrated sensor and transmitted to the cloud server through 3G, 4G, or WIFI. After receiving the data, the cloud server saves the data into the database. Then, it transfers the data to the web platform and mobile interface for tracking, visualization, and analysis when needed. On the web platform, people can examine the running status of IBEM, see historical IEQ data, and download data for further analyses. On the mobile interface, the IEQ information is provided to occupants more directly, such as the objective value and assessment score of the real-time IEQ condition and the optimization suggestion for poor IEQ factors.

Meanwhile, people can vote their feelings and satisfaction ratings regarding thermal comfort, air quality, lighting, acoustical quality, and overall environment on the mobile interface. The subjective votes will be uploaded to the cloud server and fused with the corresponding objective IEQ data "right here, right now." This is the key to how the synchronous collection of objective and subjective IEQ data is achieved.

Based on the IBEM, multiple IEQ parameters can be monitored only by one instrument, which helps solve the spatial–temporal sampling problem at a low cost. More importantly, this system connects the indoor environment and humans and breaks the information asymmetry between them. The following two sub-sections will introduce the detailed technical information of IBEM.

2.2.1 *Hardware*

The hardware of IBEM mainly refers to environmental parameter sensors, which is the foundation of IEQ data collection. In order to reduce the cost and improve the convenience of environmental monitoring, the sensor integration design is adopted in this study. However, considering the cost and size, it is impractical to include all IEQ sensors in one instrument. Therefore, after deliberations on the importance of each IEQ parameter and the need for long-term monitoring, five sensors are chosen eventually, including air temperature, relative humidity, CO_2, $PM_{2.5}$, and illuminance. Table 2.1 lists the selected sensors' information (supplier, measurement principle, range, and accuracy). The manufacturer certification periods of all sensors are more than three years.

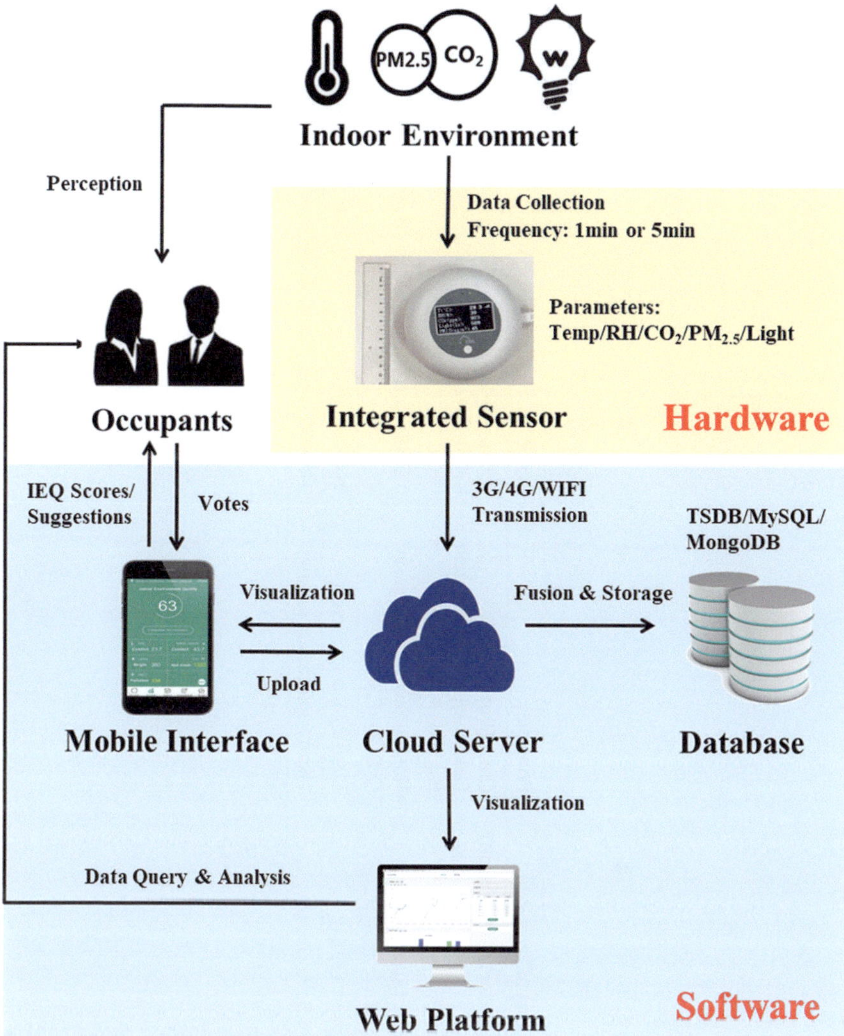

Fig. 2.1 The system architecture of IBEM

Two caveats need to be explained:

(1) Sensors of chemical pollutants, such as formaldehyde or TVOCs, are not integrated. Although these sensor costs are relatively small, the lack of specificity of low-cost TVOC sensors and the lack of accuracy at appropriate levels for formaldehyde sensors limit their applicability in this study. Furthermore, the need for long-term continuous monitoring of chemical pollutants is not specially required in Chinese standards [1–3]. The standard measurement protocol is the

Table 2.1 Information of IEQ sensors

Parameter	Product number	Sensor supplier	Principle	Range	Accuracy
Temp, RH	SHT30	Sensirion AG (Switzerland)	Complementary metal–oxide–semiconductor (CMOS) sensing	Temp: − 40 to 80 °C RH: 0–99%	Temp: ± 0.5 °C RH: ± 5%
CO_2	S8-0053	SenseAir (Sweden)	Non-dispersive infrared ray (NDIR)	0–5000 ppm	± 75 ppm
$PM_{2.5}$	PMSA003-A	Plantower (China)	Laser light scattering	0–1000 μg/m^3	± 10% @ 20–500 μg/m^3
Illuminance	BH1750FVI	ROHM Semi (Japan)	Photovoltaic effect sensing	0–50000 lx	± 5%

sampling test under certain conditions (such as closing the door, windows, air cleaning equipment, and ventilation system) rather than long-term monitoring;

(2) Noise level was not included, though acoustics is impactful on human perceptions of IEQ. It is one of the defects in hardware development, but we still have two considerations: Firstly, the noise standards [4–6] focus more on background noise and less on speech. For example, according to the Chinese national standard on noise [6], a room's acoustic performance should be assessed when the outdoor noise is maximum. Furthermore, the room should be unoccupied, the HVAC system should be running, and windows should be closed during the noise measurement. Therefore, there are no continuous sampling requirements for noise level over timeframes [7], and the one-time noise test under the most adverse condition is adequate in many cases. Secondly, unlike other IEQ parameters, there are nearly no immediate control means and technologies on the acoustic environment in most buildings. Therefore, although we could obtain the real-time noise level, the value of these data for improving operating acoustic performance is negligible. From this perspective, the continuous monitoring of noise levels seems less valuable compared to other parameters.

In addition to sensors, the hardware also includes some other functional modules: data transmission module, data storage module (backup), power module, and display module. All of them are integrated into one device, as shown in Fig. 2.2. After years of research, the integrated device has gone through several upgrades and renovations. As a result, the latest version of the device is a palm-sized round box with a diameter of 125 mm and a height of 40 mm, which can be placed on desks in real measurement scenarios.

The data transmission module can upload IEQ data to the cloud via 3G, 4G, or WIFI at the pre-defined interval (e.g., 1 min or 5 min). The 3G or 4G transmission is automatically achieved by the SIM card pre-embedded inside the device without any additional operations on site. On the other hand, the WIFI transmission requires users to connect the device to one existing WIFI in several steps, costing less than 2 min. These two signals can be used as double insurance to guarantee wireless data transmission in most cases. Furthermore, the data transmission module has a self-recovery function that the connection can be automatically recovered after a temporary network interruption. This function significantly improves the data transmission reliability in long-term monitoring.

The data storage module is mainly a backup of data storage for some cases where wireless data transmission cannot be achieved. IEQ data can be stored in a microSD card locally, which generally accommodates the data size of continuous monitoring for more than one year.

The power module includes an external power interface (USB Type-C) and one button battery for backup. The designed electrical parameters are 5 V DC and 2 A as input. Typically, the device is powered by external electricity. However, when the external power supply shuts down temporarily, the button battery can be used as a short-term power supply. Generally, the button battery can provide continuous power for about one week.

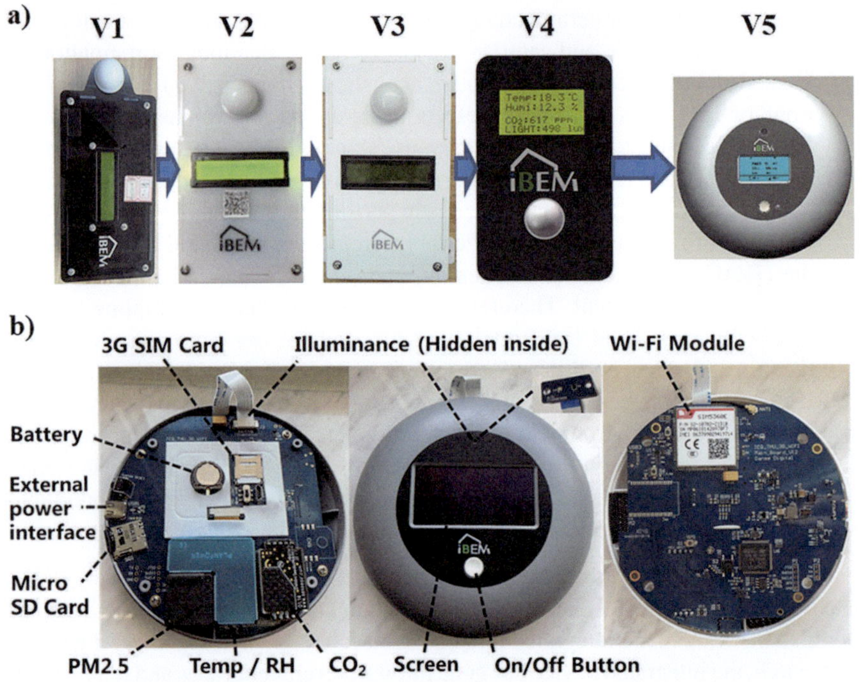

Fig. 2.2 **a** The entire development process of IBEM with different versions, and **b** the internal structure of IBEM device, including sensors, data transmission module, data storage module, power module, and display module

The display module is designed for the visualization of real-time IEQ data and the status of network connections. A LED screen is installed on the front of the device, which can be turned on by pressing the button and automatically goes off after 15 s for energy saving.

Given so many sensors and modules, it is a big challenge for the integrated device to minimize the interference, especially the influence of heat dissipation on the measurement accuracy of air temperature and relative humidity sensors. Solutions from three aspects are considered and adopted to solve this problem. Firstly, air temperature and relative humidity sensors are installed at the windward position (i.e., the edge of the mainboard closest to the external environment), so the airflow first passes through them. Secondly, the rated power of each module is controlled below 1W to reduce heat dissipation. Thirdly, the distance of primary heat sources and sensors is kept as far as possible. For example, the WIFI module is placed on the opposite side of the mainboard with the farthest distance from air temperature and relative humidity sensors. Based on the above considerations, the measurement accuracy of IBEM can be guaranteed. The performance test and validation results are further presented in Sect. 2.2.3.

2.2.2 *Software*

The software design of IBEM mainly consists of two parts: (1) Data transmission and storage (i.e., cloud server and database) and (2) Data presentation and human–computer interaction (i.e., web platform and mobile interface).

Data transmission and storage: The whole architecture is developed based on the SaaS concept, as shown in Fig. 2.3. Firstly, the Nginx load balancing technology is applied to access the multi-source data. Secondly, the Apache Tomcat 8.0 server and Linux operating system are adopted. The JavaEE Spring is used as the server programming framework. Servers provide the same API under the unified standard for calling the data conveniently. Thirdly, different databases store different data types to ensure efficient reading and writing, such as time-series database (TSDB) for dynamic IEQ data, MySQL database for static information data, and MongoDB database for subjective survey data. In addition, master–slave backup work mode is adopted in all databases, and all the data are synchronized through the intranet to ensure data security.

Data presentation and human–computer interaction: Both web platform and mobile interface are developed to build an information "bridge" between human and indoor environment, as shown in Fig. 2.4. The web platform is open to professionals and data analysts for data visualization and download. At the same time, the mobile interface is mainly designed for building occupants to assess IEQ conditions and vote on their feelings. People can easily access the mobile interface by using

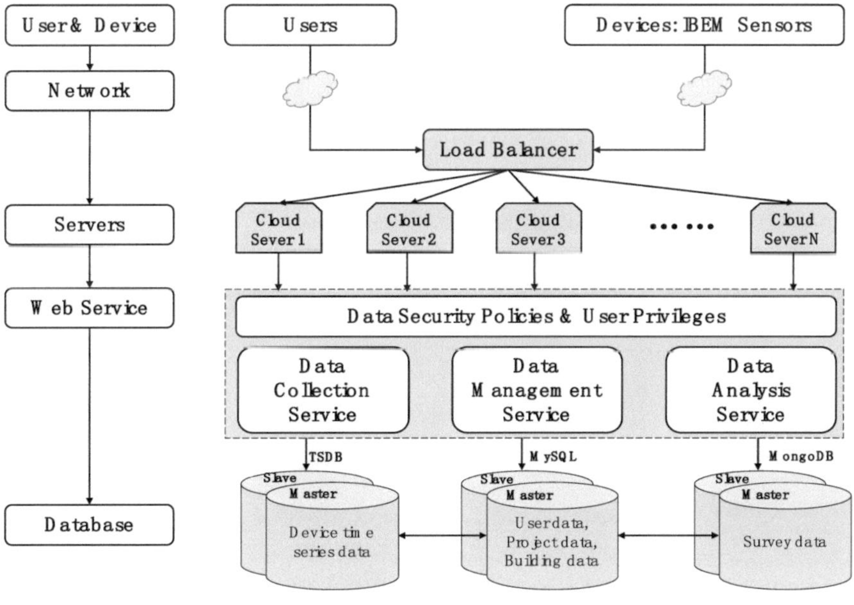

Fig. 2.3 The framework of data transmission and storage

smartphones to scan the QR code attached on the IBEM monitoring device placed nearest to them. Figure 2.4b shows several primary functions of the mobile interface, which are further illustrated as follows.

Occupants can see real-time IEQ parameters and their ratings. The rating represents an assessment of the criterion-based compliance performance of the IEQ parameter. The measurement results of specific parameters are checked against compliance criteria drawn from the relevant national or international standards (see Table 2.2) to determine the ratings. When calculating the real-time rating of one IEQ parameter, the IBEM measurements within the recent one hour are considered. That is, the rating is calculated based on the percent of measurement results within the compliance range in the last hour: The rating X represents X% of measurement results in compliance.

Fig. 2.4 Data presentation and human–computer interaction

Table 2.2 The compliance range for each IEQ parameter

Parameter	Compliance range	Source
Temperature	24–28 °C (Summer); 18–24 °C (Winter)	GB 50736-2012
Humidity	≤ 70% (Summer); No requirement (Winter)	GB 50736-2012
CO_2	≤ 1000 ppm	ASHRAE 62.1-2019
$PM_{2.5}$	≤ 35 μg/m^3	T/ASC02-2016
Illuminance	≥ 300 lx (Office)	GB 50034-2013
Sound level	≤ 45dBA (Office)	GB 50118-2010

Finally, the overall IEQ index with centesimal grade is calculated based on the weighting scheme (see Table 2.3). For thermal comfort, the weight is set to 0.30 (0.18 for temperature and 0.12 for humidity), which is generally consistent with the previous settings [8]. For IAQ, a higher weight (0.40) is put on it, compared to other studies (0.14–0.36), because Chinese people usually pay more attention to IAQ, and their satisfaction with IAQ is relatively low, according to our previous study [9]. More specifically, we set the weight of CO_2 to 0.24 and the weight of $PM_{2.5}$ to 0.16. The weight value of CO_2 concentration is higher because it can overall reflect the ventilation performance, and it is often used as a comprehensive indicator for IAQ evaluation. For lighting, the weight is set to 0.15, which is slightly lower than other studies (0.16–0.29). The reason is that the indoor lighting environment generally reaches the standard, according to the large-scale measurement of green buildings in China [10]. Thus, we lower the weight of illuminance to make the differentiation of the IEQ index more prominent and to reflect the problems of other IEQ factors to more extent. For acoustics, although the IBEM cannot measure noise level yet, this part's weight is still kept to ensure the integrity of IEQ evaluation, whose weight is lowered to 0.15, compared to other studies (0.18–0.39). Currently, we assume a maximum score for acoustics. In the future, when a noise sensor is added to the IBEM, its weight will be adjusted, and the score for acoustics will vary according to the measured data. At last, it should be claimed that the weighting scheme given by this study is just an example, and it will be revised as the study goes further.

Historical IEQ data can also be seen. However, different from the corresponding function on the web platform, the historical data query on the mobile interface only supports a recent short period (less than one day). The purpose of this design is to reduce the interface's response time and ensure a good user experience.

Table 2.3 The weighting scheme of IEQ index calculation

Thermal comfort		Air quality		Lighting	Acoustics
Temperature	Humidity	CO_2	$PM_{2.5}$	Illuminance	–
0.18	0.12	0.24	0.16	0.15	0.15

Regarding IEQ factors with poor performances, the mobile interface provides specific suggestions to help occupants make proper adjustments. For example, people are suggested to turn on air purifiers when indoor air quality is terrible.

More importantly, a voting module is developed based on the mobile interface. People can submit their feelings or satisfaction with each IEQ factor after knowing about the actual IEQ parameters. These votes are matched with objective IEQ parameters monitored by the IBEM device "right here, right now", and the combined data will be uploaded and saved in the database. In this way, the subjective and objective IEQ data can be collected simultaneously, which breaks the disadvantage of blindness to objective IEQ conditions in traditional retrospective surveys. It means that we can know how different IEQ conditions affect people's perceptions in actual buildings. This result can help us find the most satisfactory IEQ conditions for a certain space. Such information is the key to on-demand provisioning of environmental control. Another feature of this vote collection method is that it can listen to building occupants passively, instead of conducting surveys actively (which may make people annoyed). However, this approach may also cause a bias toward dissatisfaction because people are more likely to vote (complain) when they feel uncomfortable but to hold their tongues when the IEQ performance is satisfactory. This issue needs to be noticed, although more analyses are required for verification. In this study, we are not going to discuss further.

2.2.3 *Performance Test*

Before use, the IBEM devices are firstly calibrated in a testing chamber by the National Institute of Metrology (see Appendix A). In order to see the IBEM performance after authoritative calibration, this section presents a limited field evaluation in a controlled office. The performance and accuracy of the IBEM monitoring devices were tested based on the co-location strategy with several commercial sensors (reference sensors). More information of the reference sensors is presented in Table 2.4. Furthermore, these reference sensors are all within the manufacturer certification period, ensuring the accuracies are still good enough.

Co-location of both IBEM devices and their commercial equivalents was done in five locations in an office with two floors. These five locations were selected to

Table 2.4 Information the reference sensors

Parameter	Reference sensor	Range	Accuracy
Temp, RH	HOBO MX1102	Temp: 0–50 °C RH: 1–90%	Temp: ± 0.21 °C @ 0–50 °C RH: ± 2% @ 20–80%
CO_2	HOBO MX1102	0–5000 ppm	± 50 ppm
$PM_{2.5}$	TONGDY G03-PM2.5	0–600 μg/m^3	± 10 μg + 5% of the reading
Illuminance	TESTO 545	0–100000 lx	Class C according to DIN 5032-7

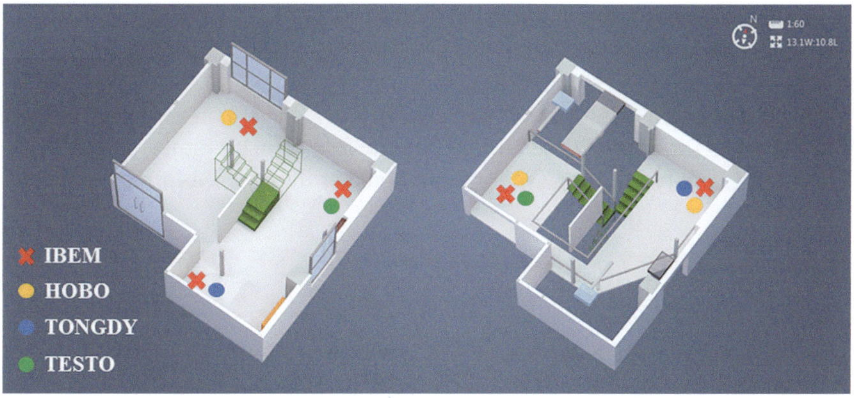

Fig. 2.5 Placement of IBEM and commercial devices for co-location testing

collect data from very different environments, which ensured the parameter range during the test was as wide as possible. Data collection occurred at 5 min intervals over two days for both commercial and IBEM devices. The averaging time is 10 min for each measurement. Figure 2.5 shows the office floor plan and the placement of devices for co-location testing. All devices were placed at desks at the height of 0.75 m.

In the office space, variable refrigerant volume air-conditioning system and humidifier were used for thermal environment (air temperature and relative humidity) control. Electric shading blinds and LED lights with adjustable luminous flux and color temperature were used for lighting control. Mechanical ventilation equipment with high efficiency filters (MERV15) was used to provide outdoor air for CO_2 concentration control, and air purifiers were used for $PM_{2.5}$ concentration control. Furthermore, to create conditions with high CO_2 and $PM_{2.5}$ concentrations during the test, dry ice and incense were used as the source of CO_2 and $PM_{2.5}$, respectively. During the experiment, the room was unoccupied, and we created considerably different IEQ conditions with air temperature varying from 15 to 27 °C, relative humidity varying from 20 to 60%, illuminance varying from 0 to 700 lx, CO_2 concentration varying from 400 to 2000 ppm, and $PM_{2.5}$ concentration varying from 10 to 200 μg/m^3. As a result, the IEQ conditions are broad enough to 'simulate' most scenarios in actual buildings.

IEQ monitoring data from the IBEM devices and their commercial counterparts were compared, and linear correlations are shown in Fig. 2.6. IBEM and HOBO data loggers revealed strong correlations for temperature and relative humidity readings throughout the test ($R^2 > 0.99$ and slopes between 1.01 and 1.02). The average deviation was 0.3 °C for temperature and 0.9% for relative humidity. CO_2 and $PM_{2.5}$ concentration measurements were also strongly correlated between IBEM devices and their counterparts ($R^2 > 0.97$ and slopes between 1.01 and 1.05), albeit with some scatter. The average deviation was only 25 ppm for CO_2 concentration and 7 μg/m^3 for $PM_{2.5}$ concentration. For illuminance, the IBEM devices showed strong correlations with the TESTO sensors ($R^2 > 0.98$ and slope = 1.01), and the average deviation was less than 11 lx. According to the test results, the measurement accuracy of IBEM devices is proved to be valid.

It should be noted that the above experiment is to test the short-term performance of IBEM under different IEQ conditions. For the long-term performance, IBEM can perform self-calibration for CO_2 concentration and illuminance in most buildings

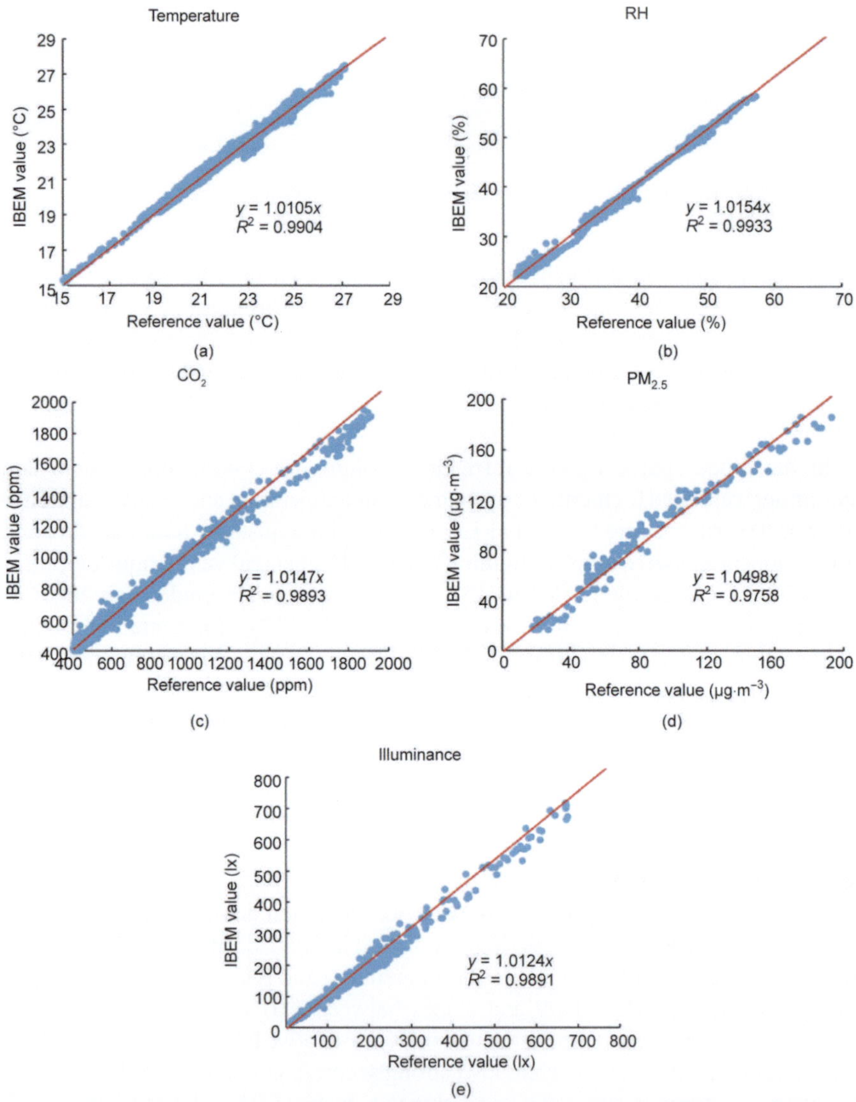

Fig. 2.6 Comparisons between IBEM devices and their commercial counterparts

because their baseline values are roughly constant over time. For example, the baseline of CO_2 concentration is 400 ppm (outdoor value). If the room is unoccupied and well-ventilated for a long time, the indoor CO_2 concentration should decrease to 400 ppm. Similarly, the baseline of illuminance is 0 lx (totally dark). If the room is unoccupied at night and all lights are closed, the indoor illuminance should be 0 lx. The self-calibration algorithm automatically compares the minimum measured value in a certain period with the baseline value and then removes the offset. However, if there are not long periods of non-occupancy in a building, it is hard to determine the IEQ parameter baseline. Therefore, the self-calibration method is not valid. We need to call back the monitoring devices for manual calibration. In terms of temperature, humidity, and $PM_{2.5}$ concentration, their baselines are not constant over time due to the influence of outdoor weather and other factors. Thus, their calibrations during the long term mainly depend on the manual way. Our study calls back each monitoring device for manual calibration and maintenance at least once a year. Furthermore, we entrusted a third-party detection institution to conduct high-frequency data acquisition tests under an extreme environment, proving the long-term operation reliability of IBEM.

2.2.4 Comparison Between the IBEM and Other Existing Tools

The large-scale applications and results from case studies have demonstrated that the IBEM can efficiently be used for IEQ monitoring, feedback, diagnosis, and improvement. However, some similar IEQ monitoring tools and systems can be found in other studies. In order to clearly show the advantages and disadvantages of different systems, this section compares their performances from five aspects: diversity of sensors, integration, extendibility, data platform, and feedback and interaction, as shown in Table 2.5.

Diversity of sensors: IBEM does not perform well, and it only includes sensors of air temperature, relative humidity, CO_2, $PM_{2.5}$, and illuminance, without acoustics and other air quality parameters. Some other systems selected more sensors for a more comprehensive IEQ monitoring. For example, SAMBA [17] also measures globe temperature, airspeed, sound pressure level, CO, formaldehyde, and TVOC, besides the above five parameters in IBEM. More extensively, the IEQ system introduced by Liu and Dai [19, 20] even monitors occupants' action in opening/closing windows and operating parameters of the mechanical ventilation system, which both have significant influences on IEQ.

Integration: Except for a small minority of studies [19, 20] directly using multiple individual market devices for IEQ monitoring, most systems, including the IBEM, develop a hardware solution that integrates a suite of sensors into a single 'box' for the low-cost and high-efficient IEQ measurement. This development of integration is also suitable for large-scale production and application.

Table 2.5 The comparison between IBEM and other intelligent IEQ monitoring tools and systems

System developer	Country	Diversity of sensors	Integration	Extendibility	Data platform	Feedback and interaction
Ali [11]	US	≥ 6	√	√	×	×
Ali [12]	US	≥ 7	√	√	√	×
Karami [13]	US	9	√	×	×	×
Sun [14]	Spain	12	√	√	×	×
Arroyo [15]	Spain	4	√	√	×	×
Perez [16]	Germany	3	√	×	√	×
Parkinson [17]	Australia	10	√	√	√	×
Moreno-Rangel [18]	UK	5	√	×	√	×
Liu/Dai [19, 20]	China	4	×	×	√	×
IBEM	China	5	√	×	√	√

Note "√" means the system has/supports the specific function; "×" means the system does not have/support the specific function

Extendibility: IBEM does not support sensor customization, and its functions are not flexible to different scenarios. However, some other IEQ monitoring systems perform better. For instance, OSBSS [11] is designed and constructed using open-source hardware and software based on the Arduino platform. All sensors are customizable, allowing for more flexibility in synchronizing a larger number of IEQ measurements cost-effectively. The IEQ system developed by Sun [14] is another example of high-level extendibility. It consists of multiple modularized sensors, and users can freely choose different sensors by a simple plug-in.

Data platform: It is well presented in IBEM and many other systems. With the development of communication and web service technology, more and more studies provide the software layer where end-users can access the data, get the suggestion, and evaluate the performance of their built environment.

Feedback and interaction: It is an essential advantage of IBEM compared to other studies. Most systems can only show the measured IEQ performance to users but cannot 'listen to users'; thus, they do not sufficiently build a bridge of information interaction between environment and human. However, the IBEM presents the real-time IEQ conditions to building occupants immediately and collects their subjective feedback at the same time. These feedbacks are essential for demand control, which can reach high human satisfaction and low energy consumption. Furthermore, the bidirectional data flow of IEQ also creates new research opportunities and more value in the field of the built environment. For example, Zhuang et al. [21, 22] found that the interactive feedback transformed building occupants' IEQ control style from a habit-driven mode to a deliberate thinking mode and induced more positive behaviors towards energy-saving, such as setting thermostats with higher temperatures in summer.

According to the above comparison results, IBEM performs better on the integration and data platform level, especially on the feedback and interaction, but worse on the diversity of sensors and extendibility.

2.3 Data Collection Method

In order to make the building performance data of different cases comparable and avoid the result deviation caused by inconsistent test methods in the process of building operating performance evaluation and analysis, this study develops a standardized data collection method of green office building operating performance, referring to relevant domestic and foreign literature and standards.

2.3.1 Building Information and Energy Use Data

Before carrying out the building operating performance test, it is necessary to collect the basic information of the building. Green building application form, self-assessment report and review report should be obtained to get the basic building information, such as the green building certification level, completion date of construction, floor area, volume fraction, building height, number of floors, number of occupants, building envelope information, and HVAC system information. Furthermore, some key design indicators of the building can be acquired (such as the designed energy consumption, energy saving rate, etc.), as well as the main green technologies adopted in the building. In the case of sufficient documents, the drawings of architectural design, construction, electrical, HVAC and other majors should also be checked to obtain more detailed building information.

Building energy consumption should be collected through the energy sub-metering system (the temporal granularity of energy consumption data can be determined according to the actual situation). For green office buildings in Severe Cold or Cold zone, heating energy consumption and non-heating energy consumption should be distinguished. For green office buildings in other climate zones, building energy consumption is divided into HVAC, lighting, office equipment, elevators, etc. If the building energy consumption includes coal, natural gas, hot water and other forms of energy, it should be converted uniformly according to the equivalent electricity method. In addition, all energy consumption data should be converted into energy consumption per floor area for comparisons with the corresponding standard or other buildings.

Particularly, if the building energy sub-metering system is not installed or out of use, the monthly energy consumption bills or payment records must be collected instead, which ensures monthly total building energy use data can be obtained at least.

Table 2.6 The criterion of determining the total number of sensors in one commercial building

Gross floor area of the building (m^2)	The total number of sensors
< 10,000	5
10,000–40,000	The total area of building/2000 (round down)
> 40,000	20

Table 2.7 The distribution percentage of sensors for room types in the commercial building

Typical types of commercial building	Main room types	Distribution percentage of sensors
Office	Open-plan office	~ 50%
	Private office	~ 40%
	Meeting room	~ 10%

2.3.2 IEQ Data

The indoor air temperature, relative humidity, illuminance, CO_2 and $PM_{2.5}$ concentrations of green office buildings need to be continuously monitored. The data collection interval should be no longer than 10 min. The total monitoring period should last for a full year. If the requirements for continuous monitoring throughout the year cannot be met, two consecutive weeks should be guaranteed for continuous monitoring in each season at least.

The total number of IEQ sensors in one office building is determined by the gross floor area, as shown in Table 2.6. The sensors should be in the main rooms of the commercial building, and Table 2.7 shows the distribution of sensors for different room types. Furthermore, as mentioned above, the selected rooms should be scattered on different floors and orientations to the greatest extent.

In each measured room, the sensor should be installed in the human-active area. In addition, the location should avoid direct sunlight and keep away from indoor heat or cold sources, such as computers and air-conditioning vents. The height of sensor placement is suggested to be 0.75 m above the floor.

2.3.3 Occupant Satisfaction Data

Building occupant satisfactions are investigated through the IBEM mobile interface (simple survey, passive collection) and electronic questionnaires (comprehensive survey, active collection). The two survey tools both have their own advantages and complement each other.

More detailed information of occupant satisfaction collection by the IBEM mobile interface can be found in Sect. 2.2.2 and Fig. 2.4.

In terms of the electronic questionnaire, the author designed a unified questionnaire to listen to users' feelings on green office building operating performances, referring to the template of CBE survey. Considering the difference of indoor environment in different seasons, the questionnaire survey should be carried out repeatedly in each season. For each building, the number of effective survey questionnaires recovered should not be less than one hundred in each season. If the total number of building occupants is small, the survey sample size can also be determined by not less than 20% of the total number of people.

2.4 Database Establishment

Based on the developed monitoring tool and method, the green office building operating performance database covering 63 buildings across different climate zones in China has been established after many years of field work.

In order to ensure the objectivity and representativeness of the building samples, the sample selection refers to the green building map issued by the Chinese Society for Urban Studies. Furthermore, the selected green office buildings are as diverse as possible in terms of green building certification level and building area, as shown in Fig. 2.7.

- Energy consumption data statistics overview

59% of the building samples have sub-metering energy consumption data (most of them are monthly sub-metering energy data, and only a few buildings have daily or hourly sub-metering energy data). 41% of the building samples only have total energy consumption data through monthly energy bills or hand-written records.

- IEQ data statistics overview

Figure 2.8 shows the on-site deployment photos of IEQ sensors in the selected green office buildings. A total of 258 sensor locations have been accumulated. At each location, IEQ parameters, including air temperature, relative humidity, CO_2, $PM_{2.5}$, and illuminance, are continuously monitored for a whole year or at least two weeks in each season.

As a result, an IEQ database with broad spatial–temporal scales and relatively fine resolutions is established. The data amount exceeds 10 GB, including more than one hundred million records of IEQ parameters.

- Occupant satisfaction data statistics overview

The database contains a total of 6040 occupant samples. Among them, 3615 samples are from the electronic questionnaire survey, and the other 2425 samples are from occupant votes via the IBEM mobile interface. Figure 2.9 shows the distribution of seasons when the occupant satisfaction data were collected, as well as the distribution of occupant gender, age, and job type.

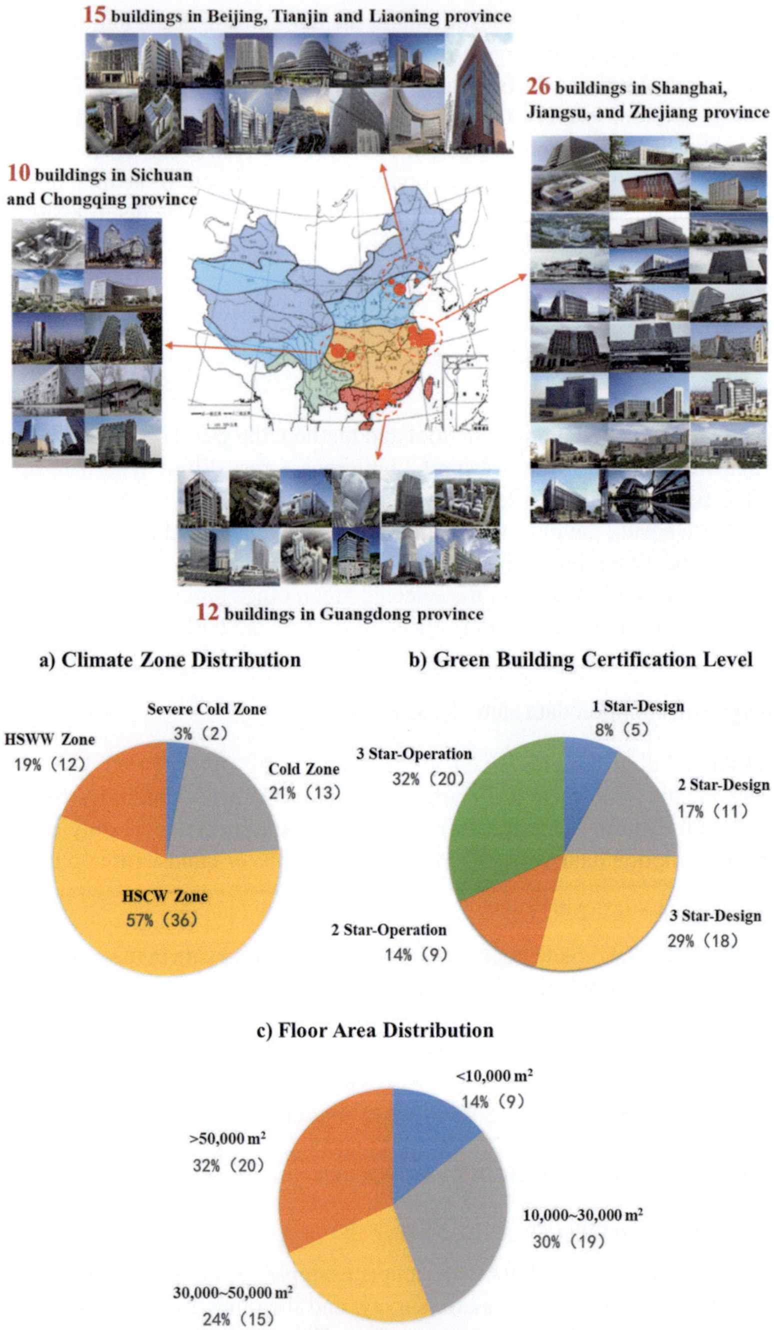

Fig. 2.7 Geographic distribution map of building samples and the basic information

Fig. 2.8 IEQ sensor deployment photos

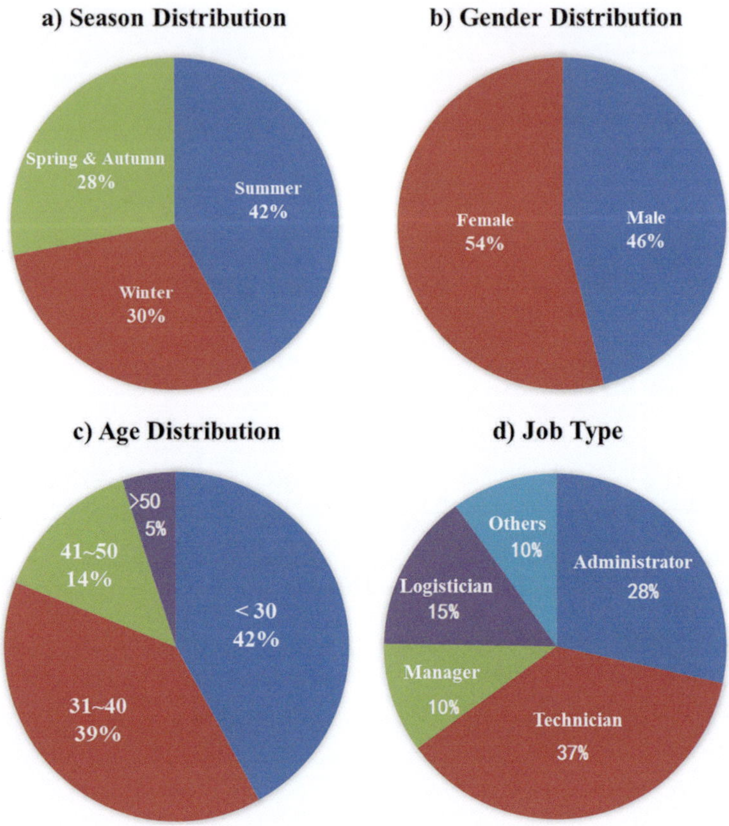

Fig. 2.9 Basic information of occupant survey data

References

1. State Administration for Market Regulation, Standardization Administration of People's Republic of China (2020) GB/T 18883–2020: indoor air quality standard. Chinese standard. Standards Press of China, Beijing. Chinese
2. Ministry of Housing and Urban–Rural Development of the People's Republic of China, State Administration for Market Regulation (2019) GB/T 50378–2019: assessment standard for green building. Chinese standard. China Architecture & Building Press, Beijing. Chinese
3. Architectural Society of China. T/ASC 02–2016 (2017) Assessment standard for healthy building. China Architecture Publishing & Media Co., Ltd., Beijing. Chinese
4. ANSI/ASA S12.2 (2019) Criteria for evaluating room noise. American National Standard. American National Standards Institute, New York
5. ISO Standard 3382-3 (2012) Acoustics-measurement of room acoustic parameters-part 3: open plan offices. ISO standard. International Organization for Standardization, Geneva
6. Ministry of Housing and Urban–Rural Development of the People's Republic of China (2010) GB 50118–2010: Code for design of sound insulation of civil buildings. China Architecture Publishing & Media Co., Ltd., Beijing. Chinese
7. Parkinson T, Parkinson A, de Dear R (2019) Continuous IEQ monitoring system: performance specifications and thermal comfort classification. Build Environ 149:241–252

8. Heinzerling D, Schiavon S, Webster T, Arens E (2013) Indoor environmental quality assessment models: a literature review and a proposed weighting and classification scheme. Build Environ 70:210–222
9. Ministry of Housing and Urban–Rural Development of the People's Republic of China, General Administration of Quality Supervision, Inspection and Quarantine (2012) GB 50736–2012: Design code for heating ventilation and air conditioning of civil buildings. China Architecture Publishing & Media Co., Ltd., Beijing. Chinese
10. ANSI/ASHRAE Standard 62.1 (2019) Ventilation for acceptable indoor air quality. American National Standard. American National Standards Institute, Atlanta
11. Ali AS, Zanzinger Z, Debose D, Stephens B (2016) Open Source Building Science Sensors (OSBSS): a low-cost Arduino-based platform for long-term indoor environmental data collection. Build Environ 100:114–126
12. Ali AS, Coté C, Heidarinejad M, Stephens B (2019) Elemental: an open-source wireless hardware and software platform for building energy and indoor environmental monitoring and control. Sensors 19:4017
13. Karami M, McMorrow GV, Wang L (2018) Continuous monitoring of indoor environmental quality using an Arduino-based data acquisition system. J Build Eng 19:412–419
14. Sun S, Zheng X, Villalba-Díez J, Ordieres-Meré J (2019) Indoor air-quality data-monitoring system: long-term monitoring benefits. Sensors 19:4157
15. Arroyo P, Herrero JL, Suárez JI, Lozano J (2019) Wireless sensor network combined with cloud computing for air quality monitoring. Sensors 19:691
16. Perez AO, Bierer B, Scholz L, Wöllenstein J, Palzer S (2018) A wireless gas sensor network to monitor indoor environmental quality in schools. Sensors (Switzerland) 18
17. Parkinson T, Parkinson A, de Dear R (2019) Continuous IEQ monitoring system: context and development. Build Environ 149:15–25
18. Moreno-Rangel A, Sharpe T, Musau F, McGill G (2018) Field evaluation of a low-cost indoor air quality monitor to quantify exposure to pollutants in residential environments. J Sens Sens Syst 7:373–388
19. Liu J, Dai X, Li X, Jia S, Pei J, Sun Y, Lai D, Shen X, Sun H, Yin H, Huang K, Tan H, Gao Y, Jian Y (2018) Indoor air quality and occupants' ventilation habits in China: seasonal measurement and long-term monitoring. Build Environ 142:119–129
20. Dai X, Liu J, Li X, Zhao L (2018) Long-term monitoring of indoor CO_2 and PM2.5 in Chinese homes: concentrations and their relationships with outdoor environments. Build Environ 144:238–247
21. Zhuang X, Wu C (2014) Saving energy when using air conditioners in offices—behavioral pattern and design indications. Energy Build 76:661–668
22. Zhuang X, Wu C (2019) The effect of interactive feedback on attitude and behavior change in setting air conditioners in the workplace. Energy Build 183:739–748

Chapter 3
Cognition of the General IEQ Characteristics of Green Office Buildings

3.1 Background

According to the literature review in Sect. 1.1.2, there are relatively comprehensive investigations on the group characteristics of green building energy consumption and occupant satisfaction, however, the research on the IEQ characteristics of green office buildings is still incomplete. Most of traditional studies only analyzed IEQ data with a small number of sensors and in a short period. Due to multiple indoor and outdoor influencing factors, the insufficient temporal and spatial distribution of IEQ data can easily lead to the randomness of the analysis results. In addition, the existing researches still lack a domestic and foreign comparison on the IEQ performance of green office buildings.

This chapter will show the actual status and comprehensive characteristics of IEQ in Chinese green office buildings, based on the long-term and massive measured data and scientific statistical analysis methods. The results will reveal the typical problems of the indoor environment that commonly exist in Chinese green office buildings green. Furthermore, the results will provide a reliable basis for the formulation or improvement of relevant IEQ standards and regulations.

3.2 Samples and Data Processing Method

In this chapter, the author selected the IEQ data of 41 green office buildings (198 IBEM sensors) from the database for the further analysis. Figure 3.1 displays the detailed locations of the selected buildings and sensors. Table 3.1 summarizes the information of all the measured buildings and their sizes. The samples covered the three main climate zones in China: Cold zone, Hot-Summer-Cold-Winter (HSCW) zone, and Hot-Summer-Warm-Winter (HSWW) zone. The amount of measured buildings and sensors in the HSCW zone was the largest (22/99), followed by the

Y. Geng, *Data-Driven Research on Characteristics of Operating Performance of Green Office Buildings*, Springer Theses, https://doi.org/10.1007/978-981-95-2329-0_3

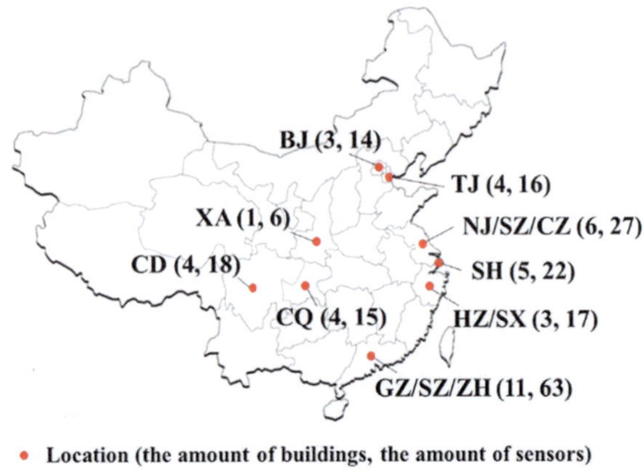

Fig. 3.1 Location of buildings and sensors (BJ: Beijing, TJ: Tianjin, XA: Xi'an, NJ: Nanjing, SZ: Suzhou, CZ: Changzhou, SH: Shanghai, HZ: Hangzhou, SX: Shaoxing, CD: Chengdu, CQ: Chongqing, GZ: Guangzhou, SZ: Shenzhen, ZH: Zhuhai)

HSWW zone (11/63) and the Cold zone (8/36). All the samples were 2-Star and 3-Star certificated green office buildings and the sample size of 3-Star buildings (34/159) was much bigger than the size of 2-Star buildings (7/39), which means the selected samples were mainly the high-level green office buildings in China. The measured samples were also divided into four groups according to the floor area of the building and the samples covered small (< 10,000 m^2), medium (10,000–30,000 m^2 and 30,000–50,000 m^2) to big (> 50,000 m^2) office buildings. Furthermore, the selected samples covered the two different building types in China [1], Type A (19/93) and Type B (22/105). Type A refers to the buildings which can use natural ventilation by opening external windows to meet indoor thermal comfort requirements and to reduce the operating time and energy consumption of HVAC system. Type B refers to the buildings which can only depend on mechanical ventilation and HVAC system to maintain indoor thermal comfort.

Due to the direct impact of outdoor climate on the IEQ conditions, it is important to know the characteristics of outdoor weather in different areas where the measured samples were located. Table 3.2 summarizes the key information of outdoor climate in these cities, according to the 2017 China Statistical Yearbook [2].

A total of 23,833,168 rows of IEQ data are included in further analyses. Since it is unavoidable that the abnormal value of IEQ data may sometimes appear during the long-term continuous measurement, data cleaning and preprocessing were conducted before the analyses. For example, CO_2 concentration data below 400 ppm[1] or above 3000 ppm have been removed. Furthermore, only the IEQ data during the working hours (9am-5 pm) on workdays was included in the analyses. The purpose of this step was to remove the IEQ data when people were not in the office buildings and to focus on the indoor environments really felt by people.

All the statistical analyses were performed using MATLAB software (Version R2014a). Shapiro–Wilk test [4] was used to examine if the IEQ data was normally

[1] According to the WMO Greenhouse Gas Bulletin [3], the global average value of atmospheric CO_2 concentration has risen to more than 400 ppm in 2016. Therefore, the indoor CO_2 concentration below 400 ppm was regarded as the abnormal data.

Table 3.1 Information of the measured samples

Categories	The amount of buildings	The amount of sensors
Climate zone		
Cold	8	36
HSCW	22	99
HSWW	11	63
Green building level		
2-Star	7	39
3-Star	34	159
Floor areas (m^2)		
< 10,000	7	31
10,000–30,000	12	50
30,000–50,000	8	41
> 50,000	14	76
Ventilation mode		
Type A (Buildings with operable windows for natural ventilation)	19	93
Type B (Buildings only with mechanical ventilation)	22	105

Table 3.2 Outdoor climate information of each city where the measured buildings were located

Climate zone	City	Mean T/RH in typical months (°C/%)				Mean $PM_{2.5}$ concentration in a year (μg/m³)
		January	April	July	October	
Cold	Beijing	-4.2/38	16.9/36	27.4/69	13.4/70	73
Cold	Tianjin	– 4.4/48	16.8/42	27.5/74	14.2/70	69
Cold	Xi'an	0.3/55	18.3/55	28.3/64	15.3/78	71
HSCW	Shanghai	4.4/72	16.7/75	29.9/75	20.8/80	45
HSCW	Nanjing	3.1/72	17.3/73	28.9/78	18.3/83	48
HSCW	Hangzhou	4.9/79	17.6/78	30.5/74	20.4/83	49
HSCW	Chengdu	5.7/80	17.6/82	25.9/86	17.9/83	63
HSCW	Chongqing	8.2/81	19.7/79	30.6/67	20.4/83	54
HSWW	Guangzhou	13.3/80	23.4/86	28.9/83	25.0/81	36

Data source 2017 China Statistical Yearbook [2]

distributed or not. In terms of significance test, t-test [5] (for normally distributed data) or Wilcoxon-Mann–Whitney test [6] (for non-normally distributed data) was used to examine the significance of difference between different seasons and climate zones. However, the data size of this study was too big and it easily led to extreme results of the significance test. All the test results showed great significance with $p <$

0.001, but some differences were small in fact. The large amount of data exaggerated the real difference between two groups of data. Thus, too much dependence on *p*-value in this study can draw misleading conclusions and the results of significant test are not shown. Instead, the median values were directly compared between different groups of data.

3.3 IEQ Characteristics in Different Seasons and Climate Zones

3.3.1 Temperature and Relative Humidity

Figure 3.2 shows the indoor thermal environment in green office buildings between different climate zones (Cold, HSCW and HSWW zone) and different seasons (Summer: from June 1st to August 31st, Winter: from December 1st to February 28th, and Transition seasons: from March 1st to May 31st & from September 1st to November 30th).

In summer, the difference of thermal environment between Cold, HSCW and HSWW zone was not significant. The median values of indoor air temperature and relative humidity in summer were 27.6 °C/51.8% in Cold zone, 26.8 °C/56.7% in HSCW zone, and 26.5 °C/54.5% in HSWW zone. The median value differences of indoor air temperature and relative humidity were within 1.1 °C and 4.9%, respectively. However, in transition seasons and winter, the difference became more significant. Firstly, the thermal environment in Cold zone was much warmer and drier than that in HSCW zone and HSWW zone. In transition seasons, the median value of indoor air temperature in Cold zone was 3.5 °C higher than that in HSCW zone and 2.0 °C higher than that in HSWW zone; in winter, the median value of indoor air temperature in Cold zone was 5.7 °C higher than that in HSCW zone and 3.7 °C higher than that in HSWW zone. In terms of relative humidity, the median value in Cold zone was 25.9% and 26.4% lower than that in HSCW zone and HSWW zone (during transition seasons); the median value in Cold zone was 18.8% and 28.8% lower than that in HSCW zone and HSWW zone (during winter). Secondly, the thermal environment in HSCW zone was slightly cooler and drier than that in HSWW zone.

In addition, the variation of thermal environment in one climate zone throughout the whole year was also summarized. In Cold zone, indoor air temperature of different seasons was relatively stable but relative humidity decreased significantly from summer to transition seasons to winter. The median value of indoor air temperature in Cold zone stayed at the range of 26.8–27.6 °C, while the median value of relative humidity in Cold zone decreased from 51.8% in summer to 26.0% in transition seasons to 14.4% in winter. In HSCW zone and HSWW zone, thermal environment had the tendency of becoming cooler and drier from summer to transition seasons to winter. The thermal environment of HSCW zone varied from 26.8 °C/

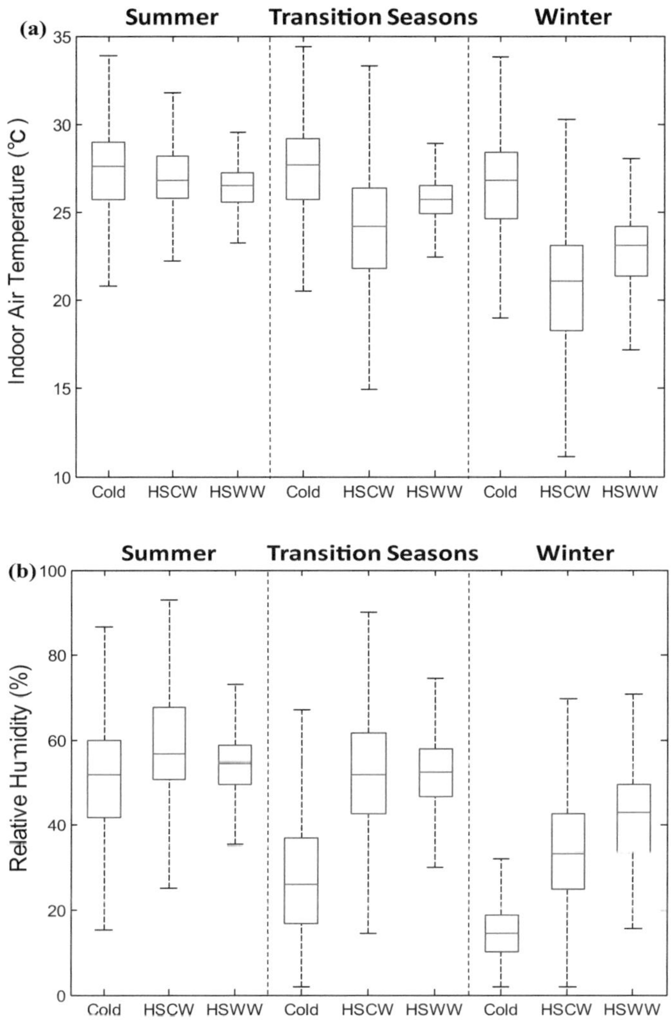

Fig. 3.2 **a** Indoor air temperature and **b** relative humidity of green office buildings in different climate zones and different seasons

56.7% in summer to 24.2 °C/51.9% in transition seasons to 21.1 °C/33.2% in winter; the thermal environment of HSWW zone varied from 26.5 °C/54.5% in summer to 25.7 °C/52.4% in transition seasons to 23.1 °C/43.1% in winter.

Figure 3.3 presents the histograms and distribution fitting curves of indoor air temperature and relative humidity. The normal or log-normal distribution was used to fit the thermal data. For indoor air temperature, the distribution of data in HSWW zone was more concentrated, compared to that in Cold zone and HSCW zone. Furthermore, in HSCW zone and HSWW zone, the distribution curves became flatter and the wave peak shifted to the left side as time went from summer to transition seasons to winter. For relative humidity, all the distributions were more dispersed compared to the distributions of air temperature, which can be found by the absolute value of density. In Cold zone, the distributions of relative humidity in transition seasons and winter were mostly concentrated on the right side, especially the distribution in winter (the majority of data were lower than 20%).

In order to evaluate the indoor thermal conditions of green buildings, the measured data were compared with the Chinese thermal comfort standard [7], as shown in Table 3.3. It should be noted that the value ranges in Chinese standard generally conform to the ASHRAE Standard 55-2013 [8] at the level of $-1 < \text{PMV} < 1$. Besides that, energy conservation is also considered in the Chinese standard, so the air temperature range in summer is limited to 24–28 °C ($0 < \text{PMV} < 1$) and the air temperature range in winter is confined to 18–24 °C ($-1 < \text{PMV} < 0$). After defining the evaluation standard, the measured thermal data were divided into several groups

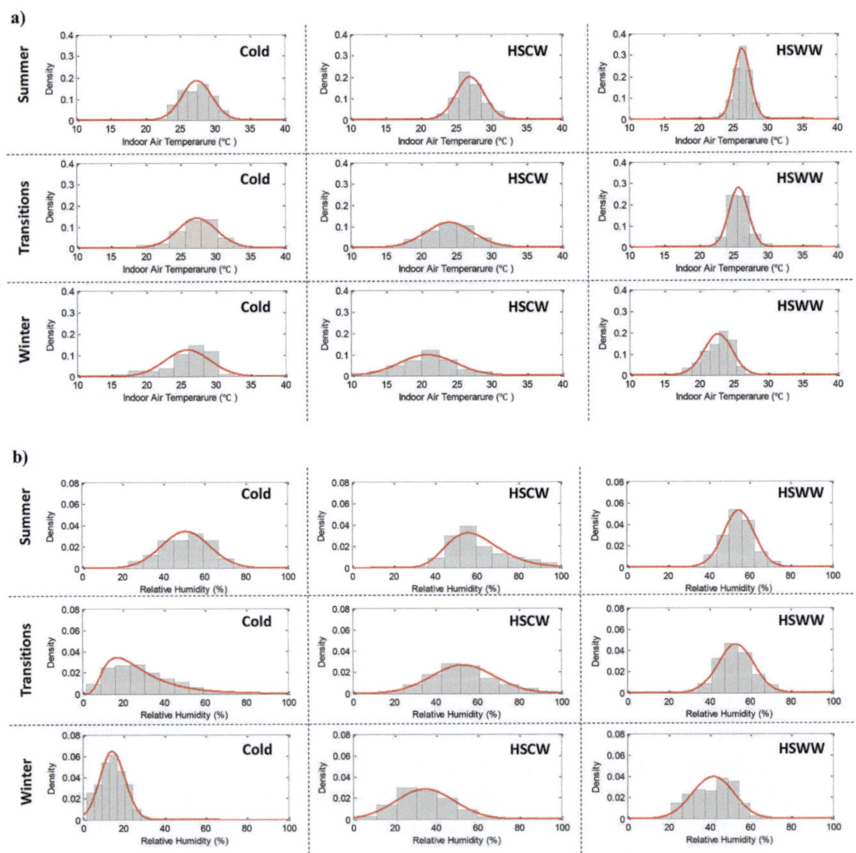

Fig. 3.3 Histograms and distribution fitting curves of **a** indoor air temperature and **b** relative humidity of green office buildings in different climate zones and different seasons

Table 3.3 Chinese thermal comfort standard of green office buildings

Season	Thermal comfort level	Air temperature (°C)	Relative humidity (%)	Air velocity (m/s)
Winter	I	22–24	≥ 30	≤ 0.2
	II	18–22	–	≤ 0.2
Summer	I	24–26	40–60	≤ 0.25
	II	26–28	≤ 70	≤ 0.3

and Fig. 3.4 shows the percentage of thermal data within each range in different climate zones and different seasons.

In summer, the qualified percentage of indoor air temperature (24–28 °C) was 61.3% in Cold zone, 67.2% in HSCW zone and 88.5% in HSWW zone. Indoor air temperature higher than 28 °C had a relatively large ratio in summer, especially in Cold zone and HSCW zone, which was called insufficient cooling problem. The percentage of insufficient cooling in summer was 34.4% in Cold zone, 28.0% in HSCW zone and 8.3% in HSWW zone.

In winter, the qualified percentage of indoor air temperature (18–24 °C) was 22.1% in Cold zone, 58.3% in HSCW zone and 65.6% in HSWW zone. It can be found that the qualified ratio was very low in Cold zone and 76.3% of data were higher than 24 °C, which was called overheating problem. Overheating problem in winter also existed in HSCW zone (18.7%) and HSWW zone (32.3%). Furthermore, in HSCW zone, there were also a large amount of air temperature data lower than 18 °C in winter, which accounted for 23.1%. This issue was called insufficient heating problem in winter.

In terms of relative humidity, most of the data were within the qualified range in summer ($\leq 70\%$), except for some data of HSCW zone. There were 21.5% of relative humidity data of HSCW zone in summer above 70%. In winter, relative humidity is not controlled according to the standard.

3.3.2 CO_2 Concentration

Figure 3.5 presents the indoor CO_2 concentration in green office buildings between different climate zones and different seasons. It can be clearly found that the indoor CO_2 concentration of green office buildings stayed at a significantly low level, no matter which climate zone the buildings located in and when the season was. The median values of CO_2 concentration were all lower than 650 ppm. The difference of CO_2 concentration was small between different climate zones and different seasons. However, there was a tendency that CO_2 concentration in HSWW zone during summer time and CO_2 concentration in Cold zone during winter time were relatively higher.

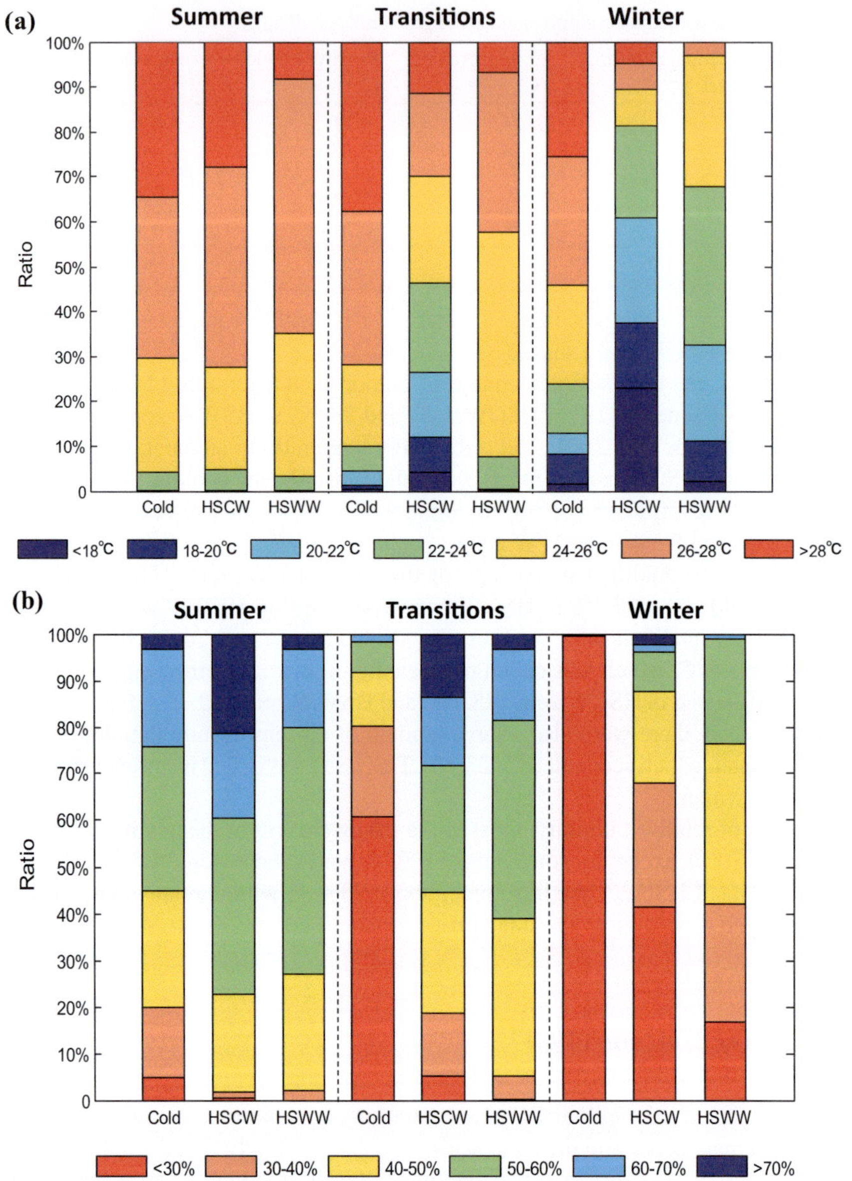

Fig. 3.4 The percentage of **a** indoor air temperature and **b** relative humidity within each range in different climate zones and different seasons

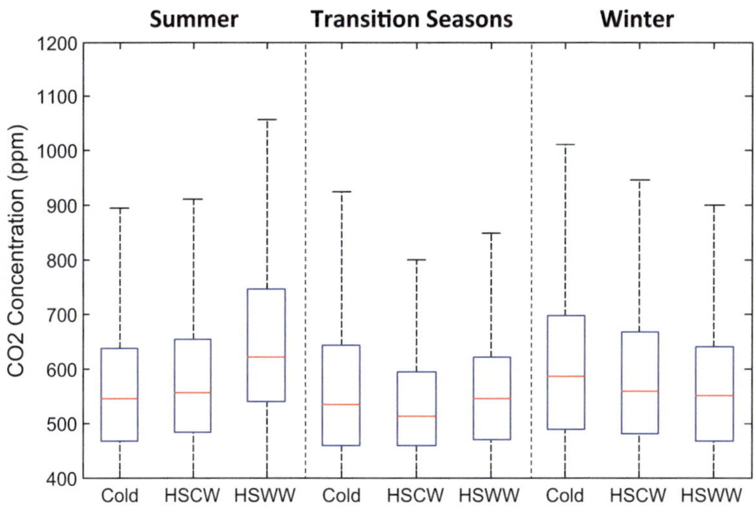

Fig. 3.5 Indoor CO_2 concentration of green office buildings in different climate zones and seasons

Figure 3.6 displays the histograms and distribution fitting curves of indoor CO_2 concentration. The distribution of CO_2 concentration had the significant characteristics of skewness and long tail, which was quite different from the normal distribution. More CO_2 concentration data concentrated on the left side, especially at the value below 600 ppm. Thus, the log-normal distribution was used to fit the CO_2 concentration data.

The threshold of indoor CO_2 concentration is 1000 ppm and the CO_2 concentration data were divided into four groups: < 600 ppm, 600–800 ppm, 800–1000 ppm and > 1000 ppm (see Fig. 3.7). Almost all the CO_2 concentration data conformed to the

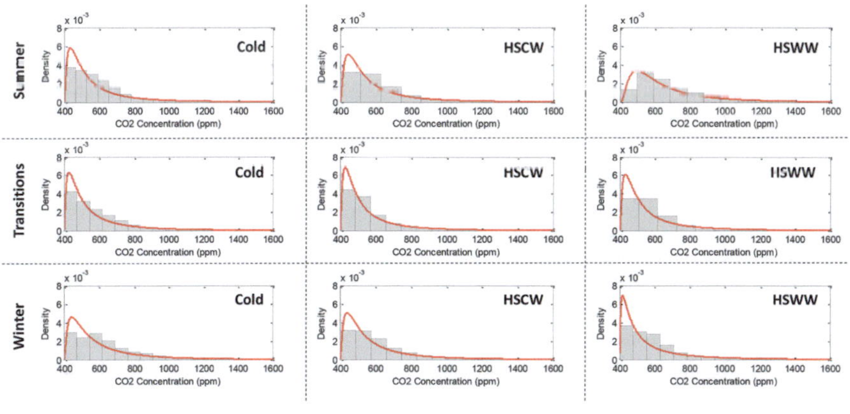

Fig. 3.6 Histograms and distribution fitting curves of indoor CO_2 concentration of green office buildings in different climate zones and different seasons

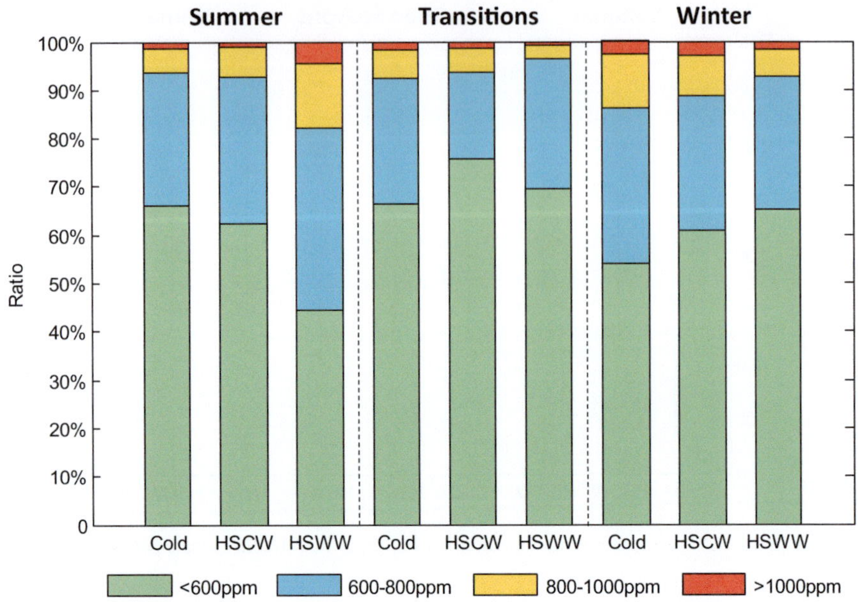

Fig. 3.7 The percentage of indoor CO_2 concentration within each range in different climate zones and different seasons

standard and that more than half of the CO_2 concentration data were below 600 ppm, except for the data in HSWW zone during summer time. These results indicate that the indoor CO_2 condition of Chinese green office buildings is generally at a good level, but this current situation also needs our rethinking if people bring in too much fresh air indoor and waste unnecessary energy accordingly. This issue will be further discussed in Sect. 3.5.2.

3.3.3 $PM_{2.5}$ Concentration

In Fig. 3.8, indoor $PM_{2.5}$ concentrations in green office buildings were compared between different climate zones and different seasons. The median values of $PM_{2.5}$ concentration varied from 12 to 28 μg/m^3. Indoor $PM_{2.5}$ concentration in Cold zone was a little higher than $PM_{2.5}$ concentration in other two climate zones. Furthermore, indoor $PM_{2.5}$ concentration during winter time was found to be relatively higher than the counterpart during summer and transition seasons. This difference was more significant in Cold zone and HSCW zone.

The distribution of $PM_{2.5}$ concentration was also left-skewed, which can be fitted by the log-normal distribution (see Fig. 3.9). This characteristic indicated that more $PM_{2.5}$ concentration data concentrated at the range of lower value. For HSWW zone, the distribution was more concentrated, compared to the counterpart in Cold zone

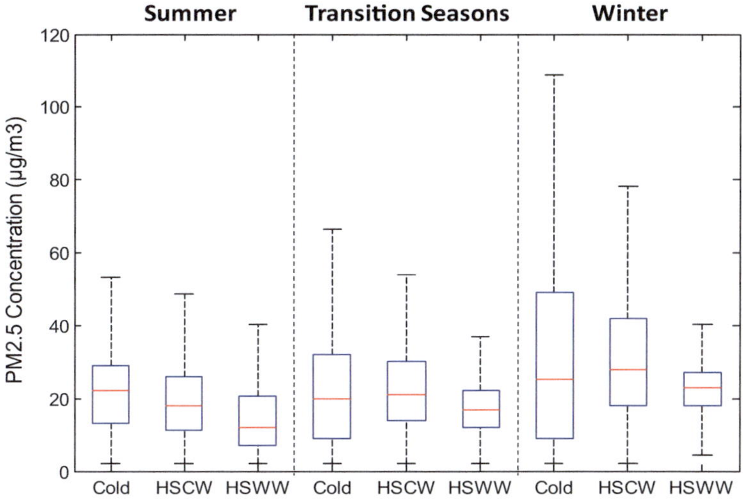

Fig. 3.8 Indoor $PM_{2.5}$ concentration of green office buildings in different climate zones and different seasons

and HSCW zone. Besides that, the location of wave peak in HSWW zone shifted to the right side as time went from summer to transition seasons to winter.

According to the Chinese Ambient Air Quality Standards (GB3095-2012) and WHO Air Quality Guidelines [9], two indicators of indoor $PM_{2.5}$ concentration were formulated in this study. The first was the upper limit of 75 μg/m^3, which is the Chinese standard value at Level II and WHO interim target I for daily mean $PM_{2.5}$ concentration. The second was the recommended value of 35 μg/m^3, which is the Chinese standard value at Level I and almost the WHO interim target III.

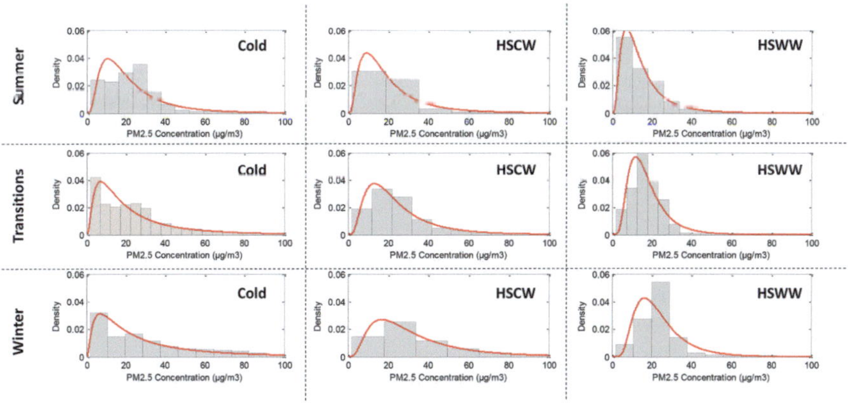

Fig. 3.9 Histograms and distribution fitting curves of indoor $PM_{2.5}$ concentration of green office buildings in different climate zones and different seasons

Figure 3.10 shows the percentage of indoor $PM_{2.5}$ concentration data within each range. Most of the indoor $PM_{2.5}$ concentration data were below the threshold of 75 μg/m^3 and the qualified rate of each data group was higher than 95%, except for the data in Cold zone during winter (87.8%) and the data in HSCW zone during winter (92.9%). These results indicate that the overall indoor $PM_{2.5}$ concentration reaches the requirement at the upper limit level. However, if 35 μg/m^3 is adopted as the guideline of indoor $PM_{2.5}$ concentration, the results will become quite different. For HSWW zone, the percentages of $PM_{2.5}$ concentration below 35 μg/m^3 in all seasons were more than 90% (92.5% in summer, 91.9% in winter and 98.0% in transition seasons). For Cold zone and HSCW zone, the ratio of $PM_{2.5}$ concentration below 35 μg/m^3 was not as much as that in HSWW zone, and there was a decrease from summer to transition seasons to winter. The percentage below recommended value of 35 μg/m^3 in Cold zone decreased from 87.6% in summer to 78.8% in transition seasons to 64.1% in winter; the percentage in HSCW zone decreased from 90.5% in summer to 82.9% in transition seasons to 64.9% in winter. Based on these results, it can be concluded that the indoor $PM_{2.5}$ level in HSWW zone is already good enough, but indoor $PM_{2.5}$ conditions in Cold zone and HSCW zone still have the space for improvements.

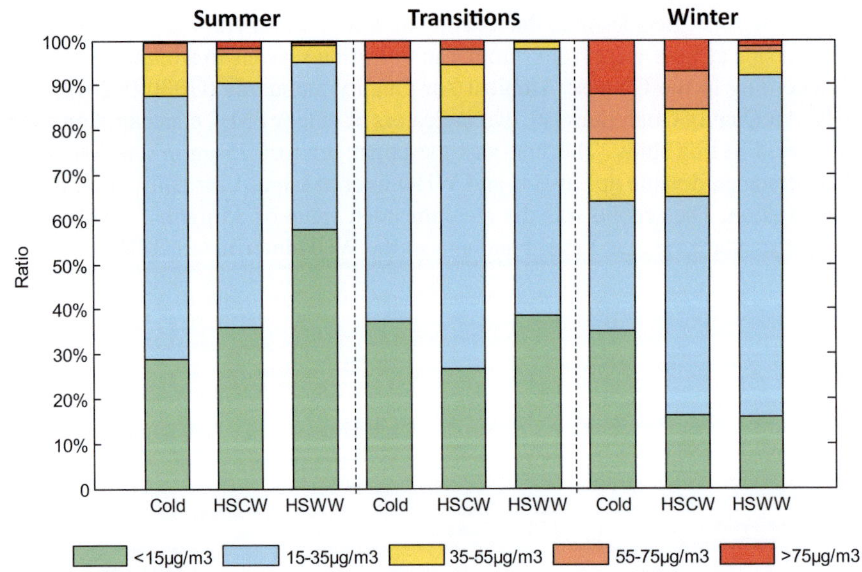

Fig. 3.10 The percentage of indoor $PM_{2.5}$ concentration within each range in different climate zones and different seasons

3.3.4 Illuminance

Indoor illuminance level is very sensitive to many factors, including the placement of sensors, the outdoor weather, the shading devices, the indoor lighting situation, etc. For example, the illuminance of the outer space near windows can sometimes be more than 10^4 lx but the maximum illuminance of the inner space is generally lower than 10^3 lx; a sudden cloud covering the sun can cause a sharp drop of indoor illuminance; an action of turning on the lights can increase the indoor illuminance directly. From the above instances, some factors are also random and hardly to be controlled, and such issues may often occur during the long-term measurement of illuminance. As a result, the illuminance data in this study varied in a very wide range and its distribution was not so regular.

Therefore, this part doesn't present the illuminance analysis diagrams like other IEQ parameters. Instead, the median values of surface illuminance in green office buildings were compared between different climate zones and different seasons (see Fig. 3.11), which can also reflect the average level of indoor lighting environment. The median values of illuminance ranged from 340 to 522 lx.

According to the Chinese standard for lighting environment [10], the thresholds for common offices and high-level offices are 300 lx and 500 lx, respectively. As shown in Fig. 3.11, the median values of illuminance in all climate zones and all seasons reached the standard for common offices and some of them even reached the standard for high-level offices. Based on these facts, it can be concluded that the indoor lighting environment of Chinese green office buildings is generally at a good level.

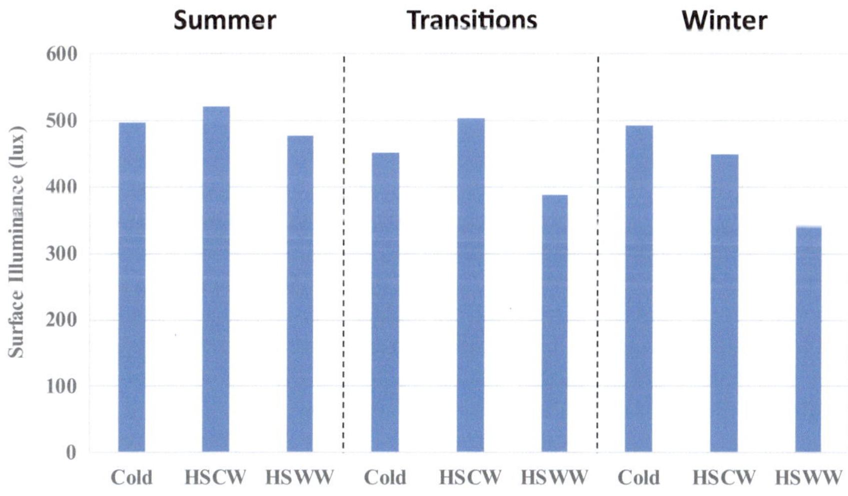

Fig. 3.11 The median values of surface illuminance in green office buildings in different climate zones and different seasons

3.4 IEQ Comparison Between China and Other Countries

In order to extend the confidence and scope of this study, the Chinese IEQ results are compared with other findings from the previous literature. Studies from the United States [11], Canada [12, 13] and Singapore [14] were selected. The measured parameters in the previous literature were similar with this study, but the sample sizes were much smaller. All of them were spot measurements or continuous measurements only for a few days. Furthermore, these previous studies didn't distinguish data in different seasons and different climate zones. Due to the big difference of data quality between this study and the previous literature, only limited comparisons are shown in this section.

Table 3.4 lists the key IEQ results in this study and others. The mean value and standard deviation of each environmental parameter were chosen as indicators. Although the IEQ measurement results from this study and others both met standard guidelines in general, several differences can be found. Thermal comfort in Chinese green office buildings had a relatively wide variation between different seasons and climate zones, whereas the thermal parameters in other countries were at a similar level, which was also close to the neutral condition. The indoor CO_2 concentration in China was generally lower than the counterpart in other countries. In terms of illuminance, all measurement results were above the lower limit of 300 lx, except for results from [12]. The above differences can be contributed to different building attributes and operation modes in different countries [15].

3.5 Discussions

3.5.1 The Relationship Between Indoor and Outdoor Parameters

Previous literature showed that outdoor weather had a big impact on the indoor environment [16, 17], thus the relationship between indoor and outdoor environmental parameters was also investigated in this study.

The monthly average values of indoor and outdoor parameters were compared in each climate zone and their correlation coefficients were calculated. Outdoor thermal data were obtained from the China Statistical Yearbook and outdoor $PM_{2.5}$ concentrations were from China Air Quality Online Monitoring and Analysis Platform (https://www.aqistudy.cn/historydata/). Outdoor CO_2 concentration remains at a relatively stable level in a year, thus it was set to be a constant value of 400 ppm in this analysis. All the detailed data are shown in Appendix B.

Figure 3.12 presents the comparison between the monthly mean value of indoor and outdoor environmental parameters and the correlation coefficients are shown in Table 3.5. A high correlation can be found between the same parameter indoors and outdoors. The variation of outdoor parameter generally led to the change of the

Table 3.4 The summary of IEQ results in this study and other studies

			Temperature (°C)	Humidity (%)	CO_2 (ppm)	$PM_{2.5}$ (μg/m^3)	Illuminance (lux)
This study	Summer	Cold	27.5 (2.2)	50.8 (11.8)	570.1 (137.5)	22.5 (13.4)	553.7 (372.5)
		HSCW	27.0 (2.0)	60.0 (13.5)	583.5 (136.4)	21.6 (24.0)	820.8 (1033.3)
		HSWW	26.4 (1.2)	54.5 (7.6)	658.0 (164.5)	15.1 (11.4)	644.1 (882.0)
	Winter	Cold	26.0 (3.2)	14.6 (6.2)	615.3 (164.2)	34.2 (32.2)	479.7 (306.4)
		HSCW	20.8 (4.1)	34.7 (14.1)	598.1 (165.0)	34.2 (24.7)	1219.9 (1898.0)
		HSWW	22.7 (2.1)	41.6 (10.0)	575.1 (144.6)	23.7 (12.9)	356.3 (156.4)
	Transitions	Cold	27.4 (2.8)	27.8 (13.9)	572.3 (160.9)	25.1 (22.8)	480.9 (351.0)
		HSCW	24.0 (3.3)	53.1 (15.0)	551.0 (137.7)	25.1 (20.0)	898.9 (1273.1)
		HSWW	25.8 (1.4)	52.8 (8.7)	560.3 (118.2)	17.5 (9.2)	448.7 (1049.3)
Ref. [11] (the United States)			23.3 (1.3)	–	733.5 (170.0)	–	448.5 (325.1)
Ref. [12] (Canada)			23.3 (1.0)	30.0 (11.0)	648.0 (97.0)	–	262.0 (252.0)
Ref. [13] (Canada)			23.0 (1.1)	35.0 (10.0)	624.0 (120.0)	–	460.0 (732.0)
Ref. [14] (Singapore)			~ 23.8 (0.5)	~ 61.9 (5.0)	~ 850 (N/A)	29.0 (N/A)	466.0 (N/A)

Note The data is mean value (standard deviation)

corresponding indoor parameter in the same direction. For example, the correlation coefficients between indoor and outdoor air temperature were 0.74 in Cold zone, 0.97 in HSCW zone and 0.83 in HSWW zone; the correlation coefficients between indoor and outdoor $PM_{2.5}$ concentration were 0.74 in Cold zone, 0.89 in HSCW zone and 0.93 in HSWW zone.

Furthermore, the cross-correlation between different parameters indoors and outdoors was also analyzed. It can be found that indoor $PM_{2.5}$ concentration had a significantly negative correlation with outdoor air temperature and their correlation coefficients were − 0.65 in Cold zone, -0.81 in HSCW zone and − 0.92 in HSWW zone. This result indicated that indoor $PM_{2.5}$ concentration was relatively high when outdoor air temperature was low. In other words, indoor $PM_{2.5}$ concentration in winter was higher than the concentration in summer and transition seasons. The analyses

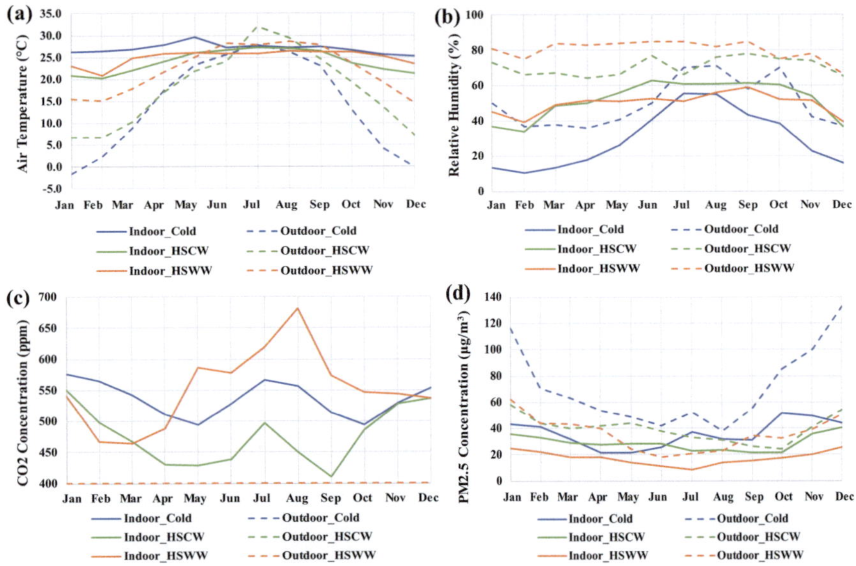

Fig. 3.12 The comparison between the monthly mean value of indoor and outdoor environmental parameters: **a** air temperature, **b** humidity, **c** CO_2 concentration, and **d** $PM_{2.5}$ concentration

Table 3.5 The matrix of correlation coefficients between each indoor and outdoor parameter

		Indoor parameters			
		T	RH	CO_2	$PM_{2.5}$
Cold zone					
Outdoor parameters	T	0.74	0.82	-0.35	-0.65
	RH	0.12	0.88	0.03	0.19
	CO_2	–	–	–	–
	$PM_{2.5}$	– 0.76	– 0.56	0.30	0.74
HSCW zone					
Outdoor parameters	T	0.97	0.90	– 0.59	– 0.81
	RH	0.34	0.52	– 0.16	– 0.36
	CO_2	–	–	–	–
	$PM_{2.5}$	– 0.64	– 0.81	0.57	0.89
HSWW zone					
Outdoor parameters	T	0.83	0.86	0.74	– 0.92
	RH	0.54	0.70	0.28	– 0.72
	CO_2	–	–	–	–
	$PM_{2.5}$	– 0.70	– 0.64	– 0.63	0.93

also found some correlations between indoor CO_2 concentration and other outdoor environmental parameters, although the strength and direction of correlation varied a lot between different climate zones.

3.5.2 Suggestions for IEQ Improvement

- Indoor thermal environment

Three possible reasons for the difference of indoor thermal environment between climate zones are listed as follows. The first reason is the heat transfer coefficient of building envelope [18, 19]. According to the design standard for energy efficiency of public buildings in China (GB 50189-2015), the heat transfer coefficient of building envelope in Cold zone is required to be lower than that in HSCW zone and HSWW zone. Thus, buildings in Cold zone have better thermal insulation performances in winter and transition seasons. The second reason is the different heating systems in different climate zones [20, 21]. In Cold zone, buildings can use hot water from urban or district heat supply network as the heat source, while buildings in the other two climate zones can only use the distributed heat sources, including air source heat pumps, ground source heat pumps, etc. Compared to the urban or district heating network, distributed heating systems have more flexibilities and can be easily adjusted to the weather, which is helpful for reducing overheating problems. The third reason may relate to the long-term thermal experience in different climate zones [22]. According to the findings of [22], people who live in the Cold zone have less acceptability for low temperature in winter due to their long-term comfortable thermal experience (neutral-to-warm) with pervasive district heating. However, people in HSCW zone and HSWW zone can forgive the cool or cold indoor environment in winter due to their long-term experience without district heating. Therefore, different thermal experiences cause different thermal expectations, and finally result in different thermal conditions in reality. The above three reasons can explain why indoor air temperature in Cold zone during winter and transition seasons can be much higher than in spite of the colder outdoor weather.

After comparing the measured thermal data with thermal comfort standard, several thermal problems were found. In order to improve thermal comfort, some suggestions are summarized as follows. Firstly, insufficient cooling in summer was found mainly in Cold zone and HSCW zone. It is suggested that, in these two climate zones during summer, the set point for indoor air temperature can be appropriately adjusted to a lower level in some hot areas. Furthermore, some personal cooling devices [23], such as small fans, can be provided to occupants to improve the local thermal environment. Secondly, insufficient heating in winter was common in HSCW zone. The operating performance of heating systems need to be improved and personal heating devices are suggested to provide to occupants in areas with relatively low air temperature, including electric radiators, heating chair cushions, etc. Thirdly, overheating in winter was pervasive throughout three climate zones, especially in Cold zone. Thus, the set point for heating is suggested to be lower in Cold zone to reduce the waste of overheating [24], though there may be a risk of decreasing occupant satisfactions on thermal environment in winter.

- CO_2 concentration

It can be seen from Fig. 3.12c that the monthly variation of indoor CO_2 concentration in a year showed a "W" shape approximately. This pattern indicated that the indoor CO_2 concentration was relatively high in summer or winter and low in transition seasons. The reason may relate to the operation of fresh air units and window-opening behaviors [25]. The outdoor weather is too hot in HSWW zone during summer and it is too cold in Cold zone during winter, thus people are less likely to open windows and building operators tend to turn off the fresh air units for the purpose of energy saving. However, during other periods, the outdoor weather is more comfortable and closer to the indoor thermal conditions, people have more possibilities to open windows for ventilation and the fresh air units run for a longer time. This inference needs further measurements in the future.

Although the indoor CO_2 condition of Chinese green office buildings is generally at a good level, the excessively low indoor CO_2 concentration indicates that too much fresh air is brought indoor and unnecessary energy may be wasted simultaneously. It is necessary to rethink the current standard and building operation on fresh air supply. Generally, what is prescribed in the existing standards is fresh air volume per capita, instead of indoor CO_2 concentration. However, the CO_2 concentration itself is more direct than fresh air volume, because the ultimate purpose of ventilation with fresh air is to maintain the indoor CO_2 concentration at a reasonable level. During the actual building operation, unorganized fresh air is unavoidable, which is also hard to predict in the designed stage. Under this situation, if building operators still run fresh air units according to the prescribed fresh air volume, the CO_2 concentration will of course be much lower than the designed value and unnecessary energy for ventilation is wasted. Therefore, it is suggested to use indoor CO_2 concentration as an indicator in the evaluation of actual air quality performance, not the fresh air volume. Furthermore, building operators should use indoor CO_2 concentration as feedback in the control of fresh air units.

- $PM_{2.5}$ concentration

The indoor $PM_{2.5}$ concentration is mainly affected by outdoor $PM_{2.5}$ concentration, infiltration rate, air filters and air cleaning devices. According to the measurements, indoor $PM_{2.5}$ conditions in HSWW zone generally conformed to the requirements of both low-level requirement (75 $\mu g/m^3$) and high-level requirement (35 $\mu g/m^3$). In Cold zone and HSCW zone, the overall indoor $PM_{2.5}$ conditions reached the low-level requirement, but there was still a certain gap between the actual value and the high-level requirement. Hence, in order to improve the indoor $PM_{2.5}$ conditions in Cold zone and HSCW zone, it is recommended to reduce natural ventilations during bad outdoor weather and to add additional air cleaning devices [26] indoor.

- Illuminance

The measurement results show that the overall illuminance of Chinese green office buildings was generally at a good level, so there is no suggestion for indoor lighting environment.

3.5.3 Limitations and Future Works

This study shows the measurement results of indoor environmental quality in Chinese green office buildings. The database can be used for other researchers to evaluate IEQ conditions of green office buildings in China and to compare with the counterpart in other countries. However, there are still some limitations of this work.

Firstly, some poor IEQ performances should be located and the reasons causing poor IEQ performances should be further studied. IEQ in naturally ventilated, mechanically ventilated and mix-mode buildings, as well as IEQ in buildings with different HVAC systems and terminals, should be divided to do more comparisons. It can help us further understand the different ways of building operation on IEQ. Besides that, human behaviors should be measured in the future and the relationship between human behaviors and IEQ should be studied as well, such as the relationship between window-opening behaviors and indoor air quality and thermal comfort. Such analyses make it possible to explore the detailed reasons behind IEQ measurement results.

Secondly, the correlation between different IEQ parameters can be also studied. For example, indoor CO_2 concentration and $PM_{2.5}$ concentration are supposed to be interrelated, because they are both influenced by the operation of fresh air units and window-opening behaviors. If some certain patterns between these parameters can be found, they can be used to predict the variation of one parameter from other parameters. Furthermore, the specific relationship between IEQ parameters under one condition can be used to examine if this condition occurs based on the measurements. For instance, the increase of indoor air temperature together with the increase of illuminance at daytime indicates more direct sunlight coming inside the room, which may be caused by the outdoor real-time weather or the occupant behavior of opening shading devices.

Thirdly, subjective survey data and energy consumption data should be added to provide a more comprehensive overview on operating performances of green office buildings. Occupant responses on indoor environment are more direct to assess whether IEQ conditions are comfortable or not. The analyses of both objective physical measurements and subjective occupant responses will be more comprehensive to evaluate IEQ [27]. In addition, energy consumption is also an important aspect of operating performances and it has a direct impact on IEQ [28]. Good IEQ performance alone doesn't mean the building operation is commendable, the amount of consumed energy must be considered simultaneously. High-quality buildings with extremely high energy consumption are not good examples. The objective is to create comfortable indoor environment at the cost of reasonable energy use instead. Therefore, it is necessary to collect the energy consumption data along with the IEQ measurement.

3.6 Summary

In this chapter, the author conducted a large-scale IEQ performance analysis of Chinese green office buildings. Based on the analysis results, three research questions have been answered:

(1) How does the IEQ performance vary between different seasons and different climate zones in China?

- The difference of indoor thermal environment between Cold, HSCW and HSWW zone was not significant in summer. However, in transition seasons and winter, thermal environment in Cold zone was much warmer and drier than the counterpart in HSCW and HSWW zone.
- Indoor CO_2 concentration was relatively high in summer or winter and low in transition seasons.
- Indoor $PM_{2.5}$ concentration in HSWW zone was generally lower than the counterpart in Cold and HSCW zone. Indoor $PM_{2.5}$ concentration in winter was relatively higher than the concentration in summer and transition seasons.

(2) Does IEQ in Chinese green office buildings show a good performance compared to standard guidelines?

- The overall thermal environment met the standard, although several thermal problems were found, such as insufficient cooling in summer, insufficient heating in winter and overheating in winter.
- The median values of indoor CO_2 concentration were far below the threshold of 1000 ppm, which indicated good indoor CO_2 conditions in Chinese green office buildings. However, the excessively low CO_2 concentration should evoke the rethinking of current standard and building operation on fresh air supply.
- The median values of indoor $PM_{2.5}$ concentration varied from 12 to 28 μg/m^3 and they generally conformed to the standard guideline. However, indoor $PM_{2.5}$ concentration needed to be further reduced in Cold and HSCW zone, in order to reach the high-level requirement.
- The median values of illuminance ranged from 340 to 522 lx. All of them reached the standard for common offices (300 lx) and some of them even reached the standard for high-level offices (500 lx).

(3) What is the difference between IEQ results in China and other countries?

- Thermal comfort in Chinese green office buildings had a relatively wide variation between different seasons and climate zones, whereas the thermal parameters in other countries were at a similar level, which was also close to the neutral condition. The indoor CO_2 concentration in China was generally lower than the counterpart in other countries. In terms of illuminance, all measurement results were above the lower limit of 300 lx, except for the result from one Canadian study.

References

1. Ministry of Housing and Urban-Rural Development (2016) Standard for energy consumption of building (GB/T 51161–2016). China Architecture & Building Press, Beijing (in Chinese)
2. China National Bureau of Statistics (2017) China statistical yearbook 2017. China Statistical Publishing House, Beijing, China (In Chinese)
3. WMO, Greenhouse Gas Bulletin (2017). Available online: https://public.wmo.int/en/resources/library/wmo-greenhouse-gas-bulletin. Accessed on 30 October 2017
4. Shapiro SS, Wilk MB (1965) An analysis of variance test for normality (complete samples). Biometrika 52:591–611
5. Efron B (1969) Student's t-test under symmetry conditions. J Am Stat Assoc 64:1278–1302
6. Neuhäuser M (2011) Wilcoxon–mann–whitney test. Springer, International encyclopedia of statistical science, pp 1656–1658
7. Ministry of Housing Urban and Rural Development (2012) Design code for heating, ventilation and air conditioning in civil buildings (GB50736-2012). China Architecture & Building Press, Beijing, China (In Chinese)
8. American National Standards Institute (ANSI)/American Society of Heating, Refrigerating and Air-Conditioning Engineers (ASHRAE) Standard 55-2013 (2013) Thermal Environmental Conditions for Human Occupancy. ASHRAE, Atlanta
9. WHO (2006) Air quality guidelines: global update 2005. Particulate Matter, Ozone, Nitrogen Dioxide and Sulfur Dioxide, World Health Organization, 2006
10. Ministry of Housing Urban and Rural Development (2013) Standard for lighting design of buildings (GB 50034–2013). China Architecture & Building Press, Beijing, China (In Chinese)
11. Choi JH, Moon J (2017) Impacts of human and spatial factors on user satisfaction in office environments. Build Environ 114:23–35
12. Newsham GR, Veitch JA, Charles KE (2008) Risk factors for dissatisfaction with the indoor environment in open-plan offices: an analysis of COPE field study data. Indoor Air 18(4):271–282. https://doi.org/10.1111/j.1600-0668.2008.00525.x
13. Newsham GR, Birt BJ, Arsenault C, Thompson AJL, Veitch JA, Mancini S, Burns GJ et al (2012) Do green buildings outperform conventional buildings? Indoor environment and energy performance in North American Offices. Research Report RR-329, National Research Council, Ottawa, Canada
14. Lee Y, Wargocki P, Chan YH, Chen L, Tham KW (2018) Indoor environmental quality, occupant satisfaction, and acute building—related health symptoms in Green Mark-certified compared with non-certified office buildings. Indoor Air
15. Geng Y, Ji W, Wang Z, Lin B, Zhu Y (2019) A review of operating performance in green buildings: energy use, indoor environmental quality and occupant satisfaction. Energy Build. 183:500–514
16. Frontczak M, Wargocki P (2011) Literature survey on how different factors influence human comfort in indoor environments. Build Environ 46:922–937

17. Chen C, Zhao B (2011) Review of relationship between indoor and outdoor particles: I/O ratio, infiltration factor and penetration factor. Atmos Environ 45:275–288
18. Manioğlu G, Yılmaz Z (2006) Economic evaluation of the building envelope and operation period of heating system in terms of thermal comfort. Energy Build 38:266–272
19. Hwang R-L, Shu S-Y (2011) Building envelope regulations on thermal comfort in glass facade buildings and energy-saving potential for PMV-based comfort control. Build Environ 46:824–834
20. Wang Z, Luo M, Geng Y, Lin B, Zhu Y (2018) A model to compare convective and radiant heating systems for intermittent space heating. Appl Energy 215:211–226
21. Lin B, Wang Z, Sun H, Zhu Y, Ouyang Q (2016) Evaluation and comparison of thermal comfort of convective and radiant heating terminals in office buildings. Build Environ 106:91–102
22. Luo M, Dear RD, Ji W, Cao B, Lin B, Ouyang Q, Zhu Y (2015) The dynamics of thermal comfort expectations: the problem, challenge and implication. Build Environ 95:231–241
23. Kim J, Bauman F, Raftery P, Arens E, Zhang H, Fierro G, Andersen M, Culler D (2019) Occupant comfort and behavior: high-resolution data from a 6-month field study of personal comfort systems with 37 real office workers. Build Environ 148:348–360
24. Liao Z, Swainson M, Dexter A (2005) On the control of heating systems in the UK. Build Environ 40:343–351
25. D'Oca S, Hong T (2014) A data-mining approach to discover patterns of window opening and closing behavior in offices. Build Environ 82:726–739
26. Bastani A, Lee C-S, Haghighat F, Flaherty C, Lakdawala N (2010) Assessing the performance of air cleaning devices–a full-scale test method. Build Environ 45:143–149
27. Fergus N, Susan R (2005) Post-occupancy evaluation and field studies of thermal comfort. Build Res Inf 33(4):338–346
28. Jaggs M, Palmer J (2000) Energy performance indoor environmental quality retrofit—a European diagnosis and decision making method for building refurbishment. Energy Build 31:97–101

Chapter 4
Joint Analysis of Energy, IEQ and Occupant Satisfaction of Green Office Buildings

4.1 Background

In the above chapter, the author reveals the IEQ characteristics of Chinese green office buildings based on the measured big data. However, building is a large complex system. The single performance analysis is not sufficient to have a comprehensive understanding of the coupling relationships between different building operating performances. Energy consumption, IEQ and occupant satisfaction are three interrelated aspects of the building operating performance. What we should pursue is the optimum balance between different building performances, rather than reducing energy consumption at the expense of indoor environmental comfort, nor blindly improving IEQ performance and occupant satisfaction without regard to energy use. Therefore, it is necessary to move from single performance analysis to comprehensive performance analysis, and explore the correlation between different building performances.

Previous researches usually lack a systematic thinking on the comprehensive building performance, and often only focus on a certain aspect of performance. Although some studies have proposed the Built Environment Efficiency (BEE) for a comprehensive evaluation of building performances, the relationship between different performances in actual buildings has not yet been clarified.

In order to fill the above research gap, this chapter will investigate the correlation between different performances in Chinese green office buildings through comparative studies. The contents mainly focus on two aspects: (1) the impact of energy saving on the IEQ performance; (2) the impact of objective IEQ condition on occupant satisfaction. In addition, this chapter will also discuss the reasons behind the results, and clarify the key factors that affect the improvement of the overall building performance with different aspects.

Y. Geng, *Data-Driven Research on Characteristics of Operating Performance of Green Office Buildings*, Springer Theses, https://doi.org/10.1007/978-981-95-2329-0_4

4.2 How Does Building Energy Conversation Affect IEQ Performance?

4.2.1 Method and Samples

In this study, 20 green office buildings from the hot-summer-cold-winter (HSCW) climate zone in China were selected. Figure 4.1 shows a map of the climate zones in China and the locations of the studied buildings. The detailed information about each building is summarized in Table 4.1, including the floor area, year of operation, green building certification level, building type, and function.

According to the Chinese green building evaluation standard, green buildings were classified into three levels (1 Star, 2 Star and 3 Star) and 3 Star is the highest level. Moreover, buildings in China are generally categorized into Type A and Type B [1]. Type A refers to public buildings with small sizes and shallow depths, which use distributed or semi-centralized air-conditioning systems and tend to rely on opening windows for fresh air. Type B refers to public buildings with large sizes and deep depths, which mostly adopt centralized air-conditioning systems and use mechanical ventilation to provide fresh air throughout the year.

After the sample selection, all buildings were firstly classified into two groups based on their energy consumption levels, and then their IEQ performances were compared. The annual total energy consumption was used as indicator to evaluate the energy performance of the green buildings. The data were mainly collected from monthly utility bills or energy-metering systems in the buildings. Among these buildings, six buildings (CQ1, SH2, SH3, SH8, SH9, and SH10) rely completely or partly on district energy stations for the cooling or heating, and three buildings have natural gas consumptions (CD2, CQ2, and SZ1). These different energy forms were finally

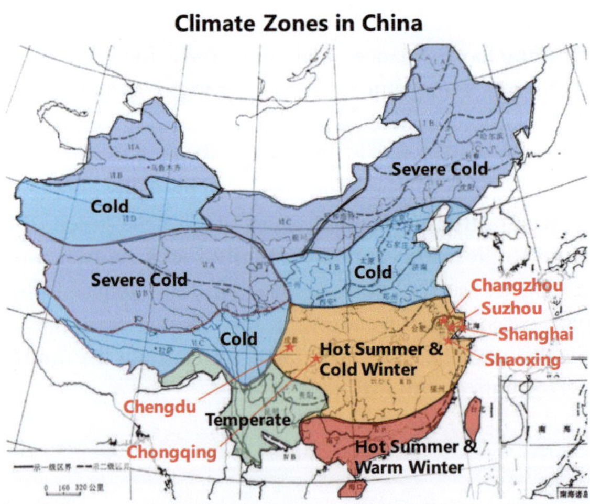

Fig. 4.1 The classification of climate zones in China and locations of the selected buildings

Table 4.1 Building information of the 20 green office buildings selected in this study

Case ID	City	Size (m^2)	Year of operation	Certification level	Building type	Function
CD1	Chengdu	12,200	2014	3 Star	A	Self-use
CD2	Chengdu	150,660	2014	2 Star	B	Rent
CQ1	Chongqing	223,000	2014	2 Star	A	Rent
CQ2	Chongqing	71,900	2013	2 Star	A	Self-use
CZ1	Changzhou	37,200	2015	3 Star	B	Rent
CZ2	Changzhou	24,878	2012	3 Star	A	Self-use
SZ1	Suzhou	70,700	2014	2 Star	B	Self-use
SZ2	Suzhou	74,900	2015	3 Star	A	Self-use
SX1	Shaoxing	39,000	2015	3 Star	B	Self-use
SH1	Shanghai	7301	2012	3 Star	A	Self-use
SH2	Shanghai	68,600	2015	3 Star	A	Self-use
SH3	Shanghai	25,500	2018	3 Star	B	Self-use
SH4	Shanghai	43,245	2015	3 Star	A	Rent
SH5	Shanghai	1900	2004	3 Star	A	Self-use
SH6	Shanghai	9992	2010	3 Star	A	Self-use
SH7	Shanghai	22,000	2013	3 Star	B	Self-use
SH8	Shanghai	95,879	2016	3 Star	B	Rent
SH9	Shanghai	89,035	2015	3 Star	B	Rent
SH10	Shanghai	55,619	2018	2 Star	B	Rent
SH11	Shanghai	89,109	2016	2 Star	B	Rent

converted into electricity, according to the electricity equivalent calculation method [2].

Furthermore, the weather normalization was conducted in order to eliminate the influence of outdoor weather on building energy consumption. According to [3], the degree days were selected as a simplification of weather data and used for weather normalization. The detailed calculation process of normalized energy is shown in Eqs. 4.1 and 4.2.

In Eq. 4.1, E_n represents the normalized energy; E_{ac} represents the actual energy; F_c represents the weather conversion factor. In Eq. 4.2, $D_{HDD18,base}$ and $D_{CDD26,base}$ represent the heating and cooling degree days using as the baselines; $D_{HDD18,ac}$ and $D_{CDD26,ac}$ represent the actual heating and cooling degree days; w_h and w_c represent the weights of heating and cooling degree days. Table 4.2 presents the real degree days of the corresponding cities in the measured year and the weather normalization results of building energy data.

$$E_n = E_{ac} \cdot F_c \tag{4.1}$$

$$F_c = w_h \cdot \frac{D_{HDD18,base}}{D_{HDD18,ac}} + w_c \cdot \frac{D_{CDD26,base}}{D_{CDD26,ac}} \tag{4.2}$$

Based on the normalized energy data, all buildings were classified into low- and high-EUI buildings. Figure 4.2 illustrates the principle of the classification and the results. Owing to the big difference of building characteristics and operation modes, the energy consumption varies greatly between Type A and Type B [4]. It can be seen that the EUI of Type A buildings ranges from 24 kWh/m^2.a to 63 kWh/m^2.a, while that of Type B buildings varies from 49 kWh/m^2.a to 110 kWh/m^2.a. It is unfair and unreasonable to mix the two types together and rank them according to their energy consumption. Instead, all 20 buildings were firstly divided into Type A and Type B. For each building type, the buildings were ranked by the actual EUI (from low to high), as shown in Fig. 4.2b. Then, the lower ranks of both types were combined as low-EUI buildings; the higher ranks were combined as high-EUI buildings. As a result, both low- and high-EUI building groups consisted of five Type A buildings and five Type B buildings respectively. This classification method distinguishes between

Table 4.2 The degree days and weather normalization results of building energy data

Case ID	City	HDD18	CDD26	Weather conversion factor	Actual energy (kWh/m^2.a)	Normalized energy (kWh/m^2.a)
CD1	Chengdu	1246	274	0.806	41.8	33.7
CD2	Chengdu	1246	274	0.806	60.3	48.6
CQ1	Chongqing	1268	182	1.039	56.2	58.4
CQ2	Chongqing	1268	182	1.039	40.1	41.7
CZ1	Changzhou	1535	283	0.735	88.3	64.9
CZ2	Changzhou	1535	283	0.735	46.2	33.9
SZ1	Suzhou	1381	274	0.778	99.0	77.0
SZ2	Suzhou	1381	274	0.778	30.8	24.0
SX1	Shaoxing	1379	306	0.722	77.6	56.1
SH1	Shanghai	1382	183	1.00	52.4	52.4
SH2	Shanghai	1382	183	1.00	58.9	58.9
SH3	Shanghai	1382	183	1.00	63.2	63.2
SH4	Shanghai	1382	183	1.00	63.0	63.0
SH5	Shanghai	1382	183	1.00	57.8	57.8
SH6	Shanghai	1382	183	1.00	56.8	56.8
SH7	Shanghai	1382	183	1.00	99.8	99.8
SH8	Shanghai	1382	183	1.00	58.9	58.9
SH9	Shanghai	1382	183	1.00	110.0	110.0
SH10	Shanghai	1382	183	1.00	101.4	101.4
SH11	Shanghai	1382	183	1.00	64.6	64.6

Type A and Type B, and guarantees the diversity and fairness of samples in two comparison groups.

The long-term IEQ data of the corresponding buildings were obtained from the database. Table 4.3 shows the number of IEQ sensors in each building. As a result, over three million sets of environmental raw data were collected for all 20 buildings.

Boxplots were mainly used to compare the measured IEQ parameters and occupant satisfactions between low- and high-EUI buildings. Before the analyses of the IEQ parameters, the data were cleaned to eliminate abnormal values of physical parameters. In addition, the measured data during non-working hours were excluded. In order to examine the significance of differences in the comparisons, Wilcoxon rank-sum test was conducted in MATLAB, and the associated p-value was calculated. However, it should be noted that large amounts of data easily result in small p-values in comparisons. When comparing the measured IEQ parameters between the low- and high-EUI buildings, the data sizes were extremely large, and all *p*-values were below 0.001. In that case, the *p*-value had little practical meaning for the assessment of the differences. Hence, the results of the significance test were not presented in the

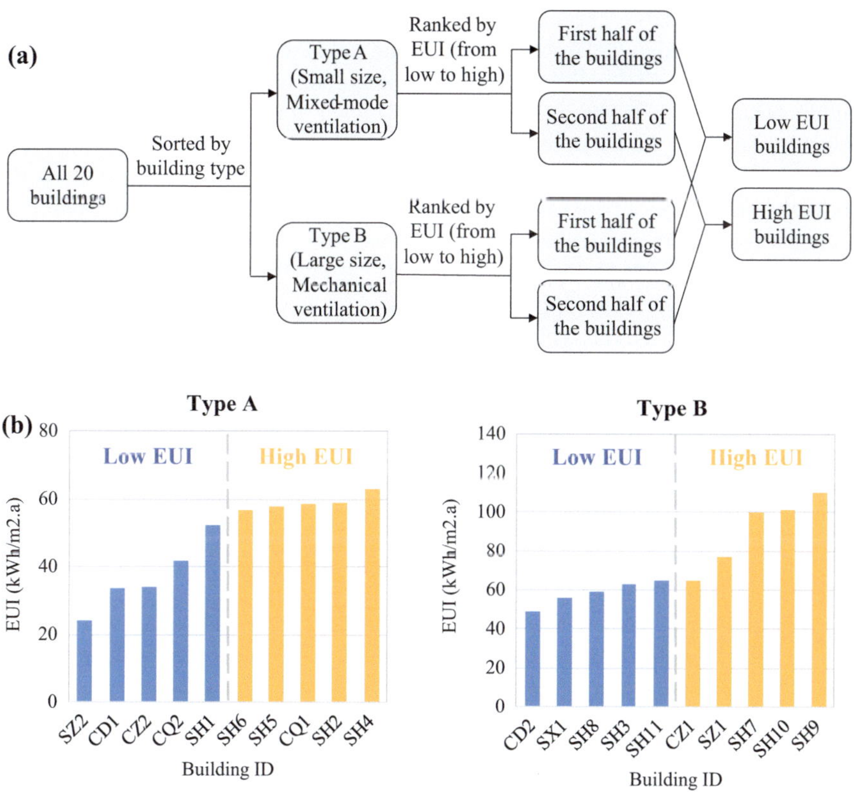

Fig. 4.2 **a** Principle of categorizing the 20 green office buildings into two groups: low-EUI buildings and high-EUI buildings; **b** rank of buildings by EUI and the classification result

Table 4.3 The total IEQ sampling number in each building

Building ID	CD1	CD2	CQ1	CQ2	CZ1	CZ2	SZ1	SZ2	SX1	SH1
Sampling number	4	6	5	4	3	3	4	6	8	7
Building ID	SH2	SH3	SH4	SH5	SH6	SH7	SH8	SH9	SH10	SH11
Sampling number	6	5	5	3	11	9	4	6	4	5

IEQ parameter comparisons. In the comparisons of occupant satisfactions, the calculated p-values exhibited good discriminant validities and were thus used to identify the differences.

4.2.2 Results

Figure 4.3 shows the comparisons of the indoor thermal environment (air temperature and relative humidity) between low- and high-EUI buildings for each season. The boxplots present the sample size, mean (blue asterisk), median (red horizontal bar), first and third quartiles, and minimum and maximum of the measured data. Compared to those of the low-EUI buildings, the mean indoor air temperature of the high-EUI buildings was 0.6 °C lower in summer, 1.2 °C higher in winter, and 0.8 °C higher in transition seasons; the mean relative humidity of the high-EUI buildings was 5.0% higher in summer, 5.4% lower in winter, and 1.1% lower in transition seasons. In addition, the difference between the indoor thermal environments of the two building groups was larger in winter than in the other seasons.

According to ASHRAE Standard 55-2013 [5], the predicted mean votes (PMV) in the low- and high-EUI buildings were calculated based on the mean air temperature and relative humidity. The typical clothing insulation was assumed to be 0.6 clo in summer, 1.1 clo in winter and 0.8 clo in transition seasons. The typical metabolic rate in the office was 1.1 met (typing). Air velocity was below 0.2 m/s, according to the spot measurements. The mean radiant was estimated to be equal to the air temperature. As a result, the PMV in the low-EUI buildings were 0.59 in summer, – 0.68 in winter, and – 0.24 in transition seasons, while the PMV in the high-EUI buildings were 0.45 in summer, – 0.45 in winter, and – 0.03 in transition seasons. Hence, the indoor thermal environment in the high-EUI buildings was closer to a neutral condition, compared to the low-EUI buildings.

The compliance rates of thermal parameters in summer and winter were calculated, according to the Chinese Design Code for Heating, Ventilation and Air Conditioning [6]. By considering the both requirements for comfort and energy conservation, the thermal comfort zone was suggested to be 24–28 °C/ $\leq$ 70% ($0 < \text{PMV} < 1$) in summer and 18–24 °C/ no specific relative humidity range ($-1 < \text{PMV} < 0$) in winter. Based on the comfort zone, all measured data were divided into three ranges (comfortable, too cold or too hot), and the ratio of each range was calculated, as

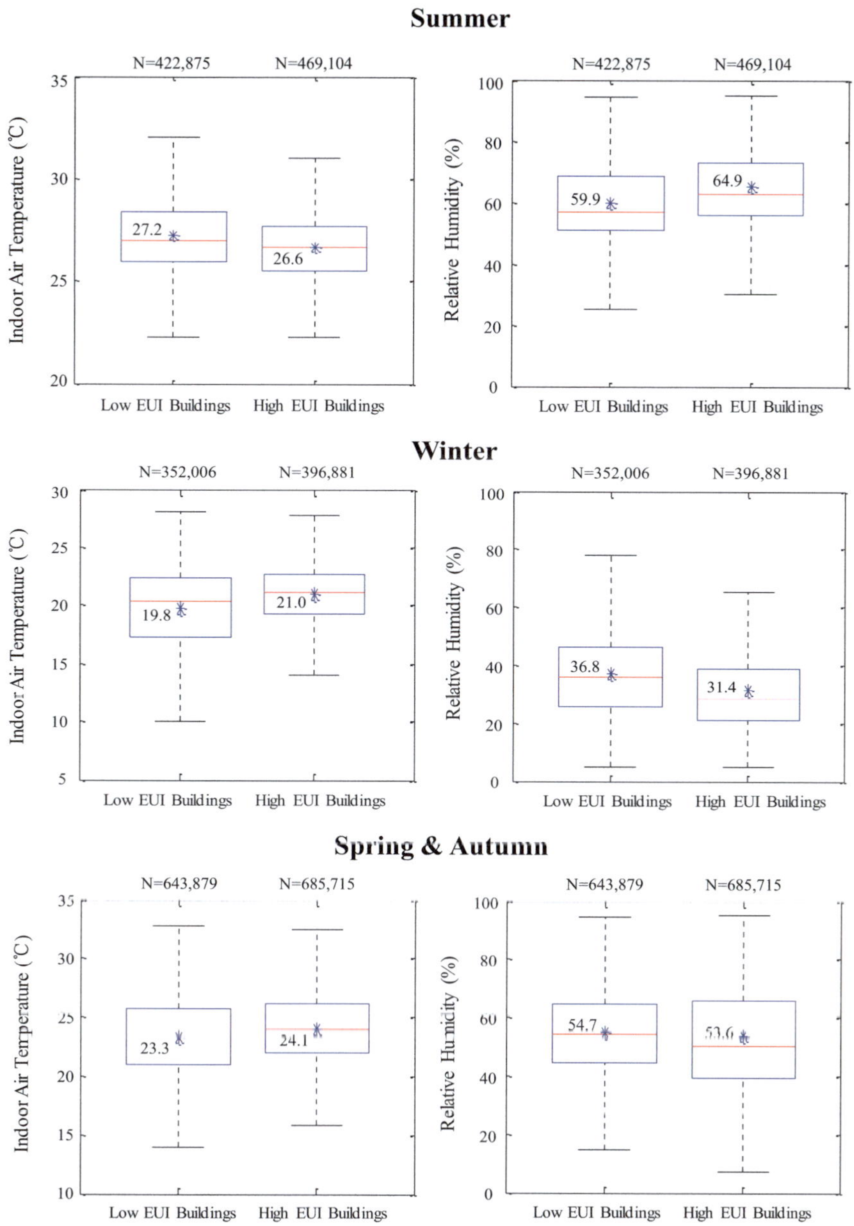

Fig. 4.3 Boxplots of indoor thermal environment (Temp. and RH) in low-EUI buildings and high-EUI buildings, which provide the size of each sample group, the mean (blue asterisk), the median (red horizontal bar), the first and third quartiles, the minimum and maximum of measured data

shown in Fig. 4.4. It can be found that the high-EUI buildings had a more comfortable indoor thermal environment than their counterparts in both summer and winter. In summer, the data ratios within the comfort zone in the low- and high-EUI buildings were 64.94% and 74.60%, respectively. The high-EUI buildings exhibited fewer 'Too hot' data than the low-EUI buildings (19.39% versus 29.96%, respectively). In winter, the data ratios achieving thermal comfort in low- and high-EUI buildings were 59.37% and 70.52%, respectively. The 'Too cold' situation was greatly improved in the high-EUI buildings compared to that in the low-EUI buildings (15.60% versus 29.39%, respectively).

Regarding the indoor air quality, the measured CO_2 and $PM_{2.5}$ concentrations of the low- and high-EUI buildings are compared in Fig. 4.5. The mean CO_2 concentrations in the low- and high-EUI buildings were 568.6 ppm and 582.1 ppm, respectively. Both values were both well below the limit of 1000 ppm in the Chinese Indoor Air Quality Standard [7]. The mean $PM_{2.5}$ concentrations in the low- and high-EUI buildings were 25.5 $\mu g/m^3$ and 30.8 $\mu g/m^3$, respectively. Although the $PM_{2.5}$ concentration was slightly higher in the high-EUI buildings, the mean $PM_{2.5}$ levels in both building groups were below the annual concentration limit of 35 $\mu g/m^3$ in the Chinese Assessment Standard for Healthy Building [8]. Figure 4.6 shows the compliance rates of the CO_2 and $PM_{2.5}$ concentrations between the low- and high-EUI buildings. The compliance rates of the CO_2 concentrations of both building groups were approximately equal (98.69% versus 98.02%, respectively), whereas the low-EUI buildings showed a slightly higher compliance rate for the $PM_{2.5}$ concentration than their counterparts (83.26% versus 75.14%, respectively). The result indicates that the indoor air quality in both building groups generally conformed to the local standard, although the high-EUI buildings performed slightly worse than the low-EUI buildings.

Figure 4.7 presents the boxplots and compliance rates of the measured illuminance. The mean values of illuminance in low- and high-EUI buildings were 488.8 lx and 519.5 lx, respectively. According to the Chinese Standard for Lighting Design

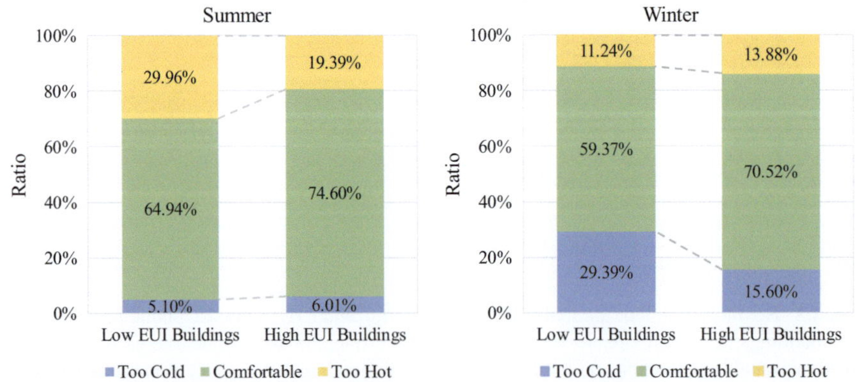

Fig. 4.4 The ratio (Comfortable, too cold or too hot) of measured thermal data in low-EUI buildings and high-EUI buildings, according to the Chinese thermal comfort standard

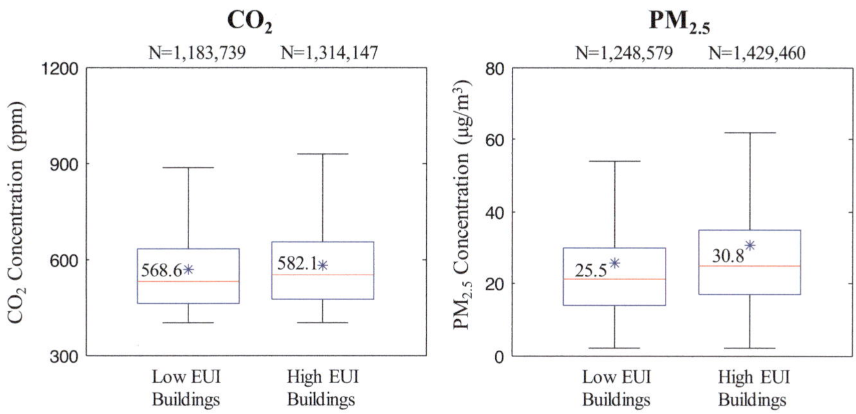

Fig. 4.5 Boxplots of indoor air quality (CO_2 and $PM_{2.5}$) in low-EUI buildings and high-EUI buildings, which provide the size of each sample group, the mean (blue asterisk), the median (red horizontal bar), the first and third quartiles, the minimum and maximum of measured data

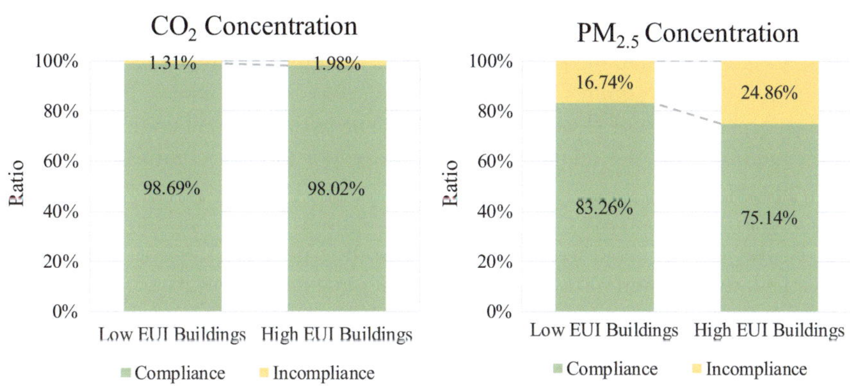

Fig. 4.6 The ratio (compliance, incompliance) of measured CO_2 and $PM_{2.5}$ concentration data in low-EUI buildings and high-EUI buildings, according to the Chinese IAQ standard

[9], the indoor illuminance for common offices should be above 300 lx. The calculated compliance rates of the illuminance in the low- and high-EUI buildings were 75.90% and 77.96%, respectively.

The objective measurements present the differences between the IEQ performances of green office buildings with different EUI levels. Regarding the thermal environment, the high-EUI buildings were significantly more comfortable than the low-EUI buildings. Thus, the relationship between thermal comfort and energy use was further discussed, as shown in Fig. 4.8. For each building, the thermal data outside the comfort zone were divided into two groups, which had different correlation patterns with energy use. The ratio of insufficient cooling or heating represented the average ratio of thermal parameters, which were above the comfort zone in

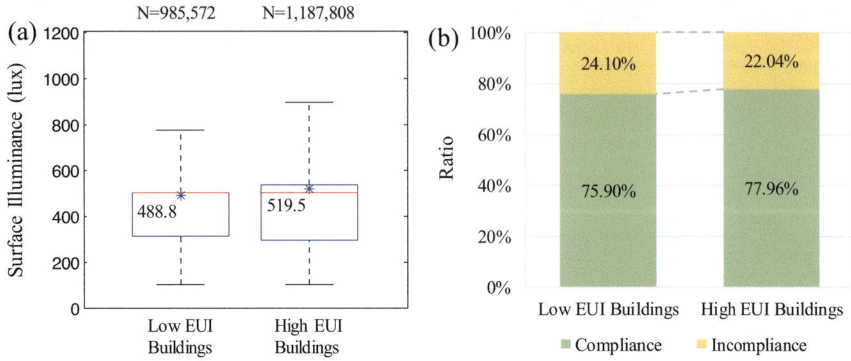

Fig. 4.7 **a** Boxplots of indoor visual environment (illuminance) in low-EUI buildings and high-EUI buildings, which provide the size of each sample group, the mean (blue asterisk), the median (red horizontal bar), the first and third quartiles, the minimum and maximum of measured data; **b** the ratio (compliance, incompliance) of measured illuminance data in low-EUI buildings and high-EUI buildings, according to the Chinese lighting design standard

summer and below the comfort zone in winter. Conversely, the ratio of overcooling or overheating represented the average ratio of thermal parameters which were below the comfort zone in summer and above the comfort zone in winter. Furthermore, building type A and type B were distinguished due to the different operation modes, which might impact the relationship between indoor thermal comfort and energy consumption.

Along with the increase of the energy consumption in both building type A and type B, the ratio of insufficient cooling or heating decreased, while overcooling or overheating occurred more often. This result indicates that green office buildings consuming more energy indeed helped guarantee the minimal requirement of thermal comfort, however, they also tended to bring about more unnecessary energy waste problems.

For building type A, the positive effect on thermal comfort (the decrease of insufficient cooling or heating) caused by high energy use was more significant than the negative effect (the increase of overcooling or overheating), which means energy was more effectively consumed in the improvement of the thermal environment. Nevertheless, for building type B, the positive effect was not evident and even smaller than the negative effect. The regression of the insufficient cooling or heating and energy use had a small value of R^2 (0.37), and its slope was flatter, compared to the regression of overcooling or overheating and energy use. This result illustrates the increase of energy consumption in building type B did not significantly reduce the insufficient cooling or heating problem, but caused a higher energy waste on overcooling or overheating.

The above results can be explained by the different operation modes and building sizes of type A and type B. According to the classification criteria of building type A and type B, type B could hardly use natural ventilation to achieve indoor thermal comfort and its size was generally bigger than that of type A. As a result, thermal

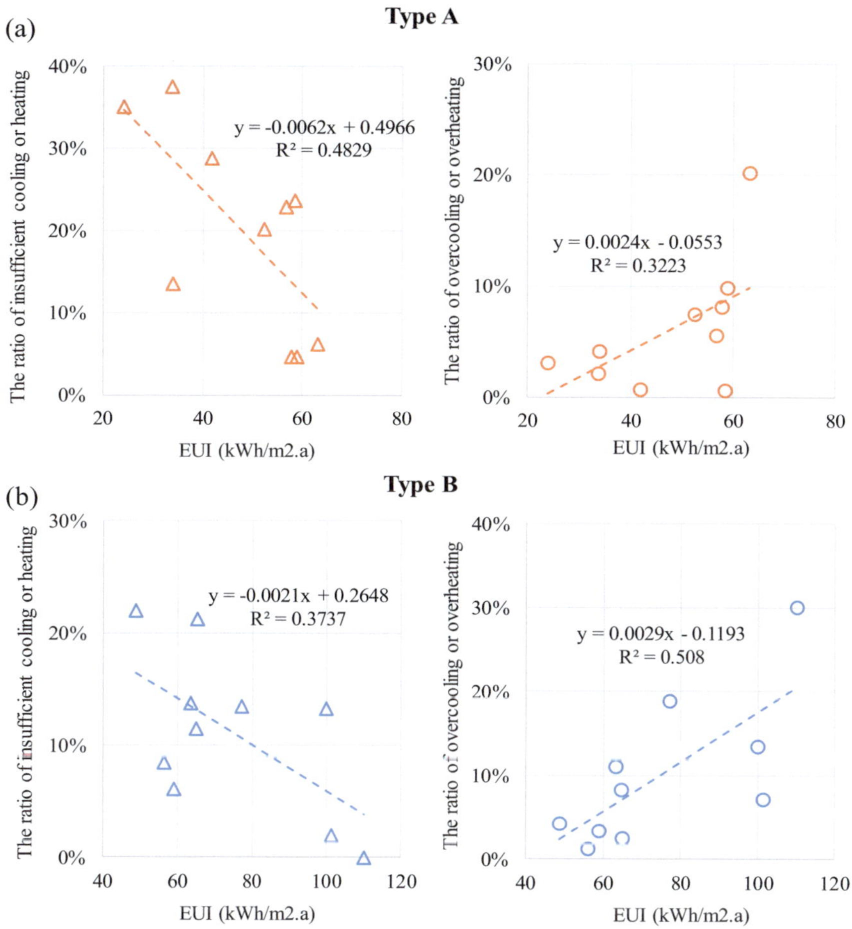

Fig. 4.8 The relationship between thermal environment and energy use

comfort in building type B was usually more difficult to maintain and there were more 'dead zones' for air-conditioning, due to the deep plan and complex space division. Even though more energy was consumed, it did little to help reach the minimal thermal requirements of "dead zones". To make matters worse, these "dead zones" might also result in a thermal imbalance in a large space. A central HVAC system without proper commissioning can increase the total amount of cold or heat supply in order to compensate for the dead zones. However, such practice easily leads to overcooling or overheating in other zones. Therefore, the increase of energy consumption in building type B improved the insufficient cooling or heating problem only slightly, and caused more energy wastes owing to overcooling or overheating.

4.2.3 Conclusion

In this study, the relationship between energy use and IEQ in green office buildings was investigated. Based on investigations of energy use, long-term environmental measurements, and occupant satisfaction surveys in twenty green office buildings from the HSCW zone in China, comparative analyses of the IEQ between low- and high-EUI buildings were performed.

The analyzing results from the measured parameters show that the difference of IEQ performance between low- and high-EUI buildings was mostly reflected in thermal environment. Compared to that of the low-EUI buildings, the indoor thermal environment in the high-EUI buildings was closer to a neutral condition, and the compliance rate of the thermal parameters was 9.66% higher in summer and 11.15% higher in winter. Regarding air quality and visual environment, the difference between the two groups was not so significant and the mean levels conformed to the local standard. However, the subjective surveys led to an inconsistent conclusion that the occupants in high-EUI buildings were less satisfied with the IEQ (except for visual environment) than their counterparts in low-EUI buildings. The mean satisfaction scores of thermal comfort, air quality and overall environment in high-EUI buildings were 0.04, 0.11 and 0.17 lower than those in low-EUI buildings, respectively. However, the differences in the thermal comfort and air quality were not statistically significant ($P > 0.05$). The possible underlying reason for these results was discussed and the perceived environmental control was investigated. The mean satisfaction and self-reported frequency of environmental control were significantly lower in high-EUI buildings, compared to low-EUI buildings ($P < 0.001$). This finding can partly account for the low perceived comfort in high-EUI buildings despite the good objective IEQ parameters.

Furthermore, the detailed relationship between energy use and thermal environment was studied. Both positive and negative effects of increasing EUI on the thermal environment were observed. While green office buildings consuming more energy indeed helped guarantee the minimal requirements of thermal comfort, they also tended to bring about more unnecessary energy wastes on overcooling or overheating. Besides, building type A was found to use the increasing energy more effectively on the improvement of thermal environment than building type B.

4.3 How Does Objective IEQ Performance Affect Occupant Satisfaction?

Based on the IBEM mobile interface, we have collected 2425 building occupants' satisfaction votes, which were matched with environmental parameters "right here, right now". Occupant votes use a 7-point scale with "+3" indicating "very satisfied" and "-3" indicating "very dissatisfied". Distributions of occupant satisfaction under different IEQ conditions were analyzed and compared. The followings show

how indoor air temperature, CO_2 concentration and illuminance affect occupant satisfaction, respectively.

4.3.1 The Effect of Air Temperature on Thermal Satisfaction

Differences in climate and HVAC systems between northern and southern China can influence thermal experiences and preferences. Thus, the collected occupant votes were split into two parts: Occupants from the Severe Cold and Cold zone were gathered to represent the northern group, while occupants from the HSCW and HSWW zone represented the southern group.

Thermal satisfaction votes of these two groups under different indoor air temperature conditions during summer and winter were compared, as shown in Fig. 4.9. The statistical test (Wilcoxon-Mann–Whitney test) was conducted, and p-values between the two data groups were calculated. If the difference is statistically significant, it is marked with an asterisk ($P < 0.05$) or two asterisks ($P < 0.01$). Furthermore, the percentage of satisfaction (i.e., votes from "0" to "+3") was calculated at each temperature partition, which was also compared to the predicted result according to the PMV-PPD model. The predicted percentage of satisfaction was calculated based on several assumptions: (1) The mean radiant temperature equaled to the air temperature; (2) The airspeed was below 0.2 m/s; (3) The metabolic rate of humans was 1.0 met; (4) The clothing insulation in summer and winter was 0.5 and 1.0 clo, respectively.

During summer, 24–26 °C was the most satisfactory thermal condition for both northern and southern groups. However, the northern group had a weaker acceptance of cooler conditions than the southern group. When indoor air temperature was below 22 °C, thermal satisfaction of the northern group dropped more sharply than that of the southern group ($P < 0.05$ when the temperature was within 20–22 °C; $P < 0.01$ when the temperature was below 20 °C). Compared to the PMV-PPD model, the percentage of satisfaction from the northern group was also below the model-based prediction when air temperature went below 22 °C. This result indicates that buildings in northern China should avoid the overcooling problem in summer for thermal environment control. Otherwise, occupant satisfaction will decrease dramatically, along with the problem of energy waste.

In winter, thermal satisfaction from the southern group varied more smoothly with air temperature, and their relationship was consistent with the PMV-PPD model. However, due to living in an environment with central heating for years, occupants from northern China were pickier about thermal conditions in winter. As long as the air temperature was not within the range of 20–26 °C, the actual thermal satisfaction from the northern group would decrease significantly, which was much below the southern counterpart, and PMV-PPD prediction results in the same condition ($P < 0.05$ when the temperature was below 18 °C; $P < 0.01$ when the temperature was within 18–20 °C and above 26 °C). Such a result indicates that building occupants in the north have a narrower range of adaptations to the thermal environment in

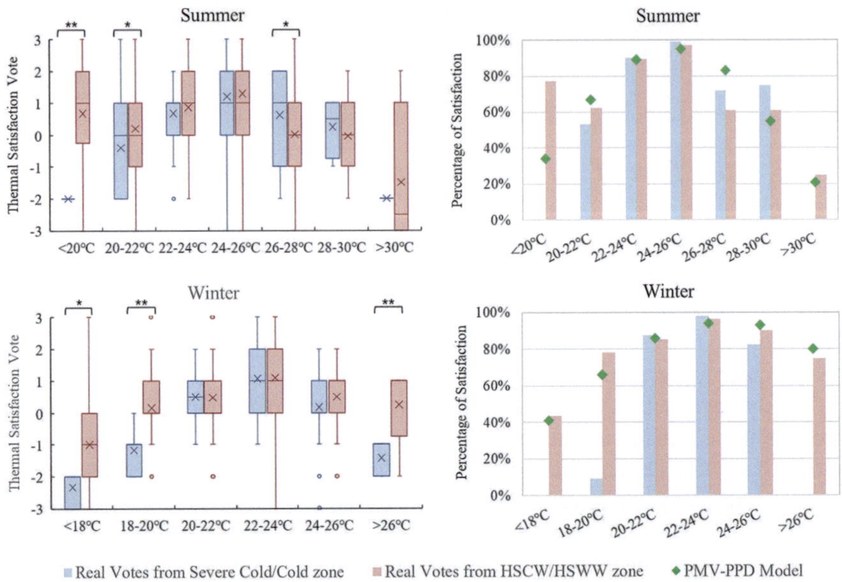

Data group in Summer	<20°C		20-22°C		22-24°C		24-26°C		26-28°C		28-30°C		>30°C	
	North	South	North	South	North	South	North	South	North	South	North	South	North	South
The number of samples N	5	42	87	128	90	116	236	342	78	90	20	45	4	19

Data group in Winter	<18°C		18-20°C		20-22°C		22-24°C		24-26°C		>26°C	
	North	South	North	South	North	South	North	South	North	South	North	South
The number of samples N	16	44	18	51	82	151	280	323	58	68	12	20

Fig. 4.9 The relationship between indoor air temperature and occupant thermal satisfaction. *Notes* "Blue" represents the northern group (severe cold/cold zone), "Red" represents the southern group (HSCW/HSWW zone), and "Green" represents the predicted result of the PMV-PPD model; *Significance test* $^{*}P < 0.05$ and $^{**}P < 0.01$

winter. Therefore, it is challenging for northern China to achieve high satisfaction in temperature control during winter.

4.3.2 The Effect of CO_2 Concentration on IAQ Satisfaction

Before studying the relationship between CO_2 concentration and occupant satisfaction of air quality, we should first distinguish how buildings are ventilated. Buildings in China are generally categorized into Type A and B, according to different ventilation types. Type A refers to buildings whose windows can be opened for natural ventilation, while type B refers to buildings using mechanical ventilation throughout the year.

Figure 4.10 shows the occupant votes of indoor air satisfaction at different CO_2 concentrations and the comparison between building types A and B, tested by the Wilcoxon-Mann-Whitney test. It is found that the average indoor air quality satisfaction at different levels of CO_2 concentration varied in a small range, no matter in building type A or B. The average satisfaction score was always above zero at any level of CO_2 concentration. However, the votes extended to "-3" for all data groups, which suggested that a substantial minority always considered air quality unacceptable, regardless of the objective IAQ performance. The result indicates that people are not very sensitive to CO_2 concentration, and it is hard to increase subjective satisfaction by improving objective performances. From the perspective of energy-saving, it is unnecessary to control CO_2 concentration too low, as long as the limit of 1000 ppm is not exceeded. However, occupants in building type A seemed to have a lower acceptance for relatively high CO_2 concentrations than occupants in building type B. When the CO_2 concentration was above 800 ppm, the occupant satisfaction of air quality in building type A was significantly lower than in building type B ($P < 0.01$).

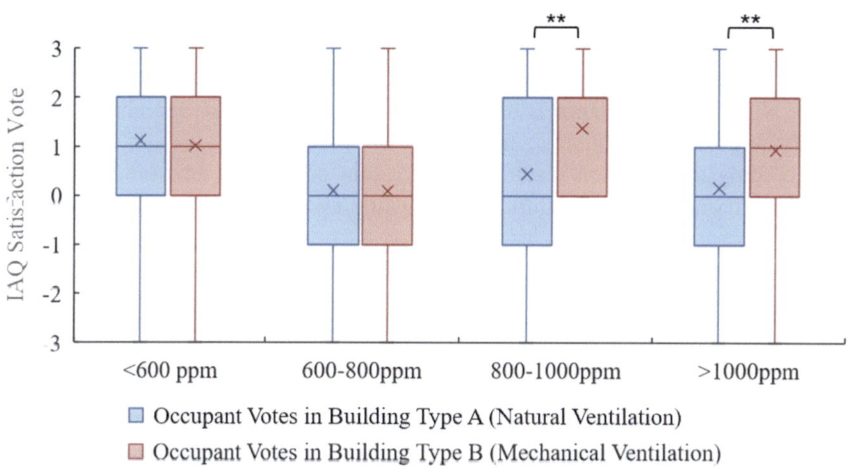

Data group	<600ppm		600-800ppm		800-1000ppm		>1000ppm	
	Type A	Type B	Type A	Type B	Type A	Type B	Type A	Type B
The number of samples N	593	282	462	334	258	257	131	108

Fig. 4.10 The relationship between indoor CO_2 concentration and occupant air quality satisfaction. *Notes* "Blue" represents occupant votes from building type A, and "Red" represents occupant votes from building type B; Significance test: $^{**}P < 0.01$

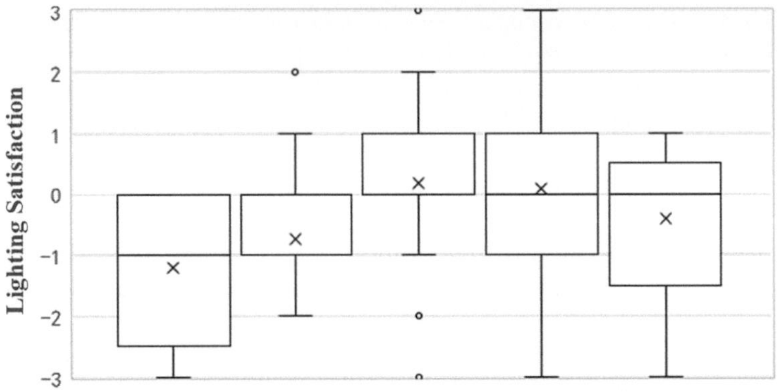

Fig. 4.11 The relationship between indoor illuminance and lighting satisfaction

4.3.3 The Effect of Illuminance on Lighting Satisfaction

Figure 4.11 shows the correlation between the lighting environment and occupant satisfaction. When the illuminance is 300–500 lx, the lighting satisfaction is the highest; when the illuminance is lower than 300 lx, the satisfaction drops significantly; when the illuminance is higher than 500 lx, the satisfaction also decreases though the decline is relatively small.

According to the above results, the key point of lighting environment control should be to ensure sufficient illuminance while avoiding the occurrence of direct glare.

4.4 Summary

This chapter breaks through the shortcomings of previous studies that only focus on a single aspect of building performance, reveals the correlation between different performances in actual green office buildings, and further clarifies the key factors to achieve the comprehensive improvement of building energy efficiency, IEQ and occupant satisfaction.

References

1. Ministry of Housing Urban and Rural Development (2016) Standard for Energy Consumption of Building (GB/T51161-2016). China Architecture & Building Press, Beijing, China (In Chinese)
2. Yi J, Xiu Y (2010) Electricity equivalent conversion method in energy analysis. Energy China 32:5–11

3. Li Z, Li H, Yu S, Zhu Y (2010) Equivalent energy consumption of building energy efficiency. J Tongji Univ 38:353–357
4. Lin BR, Liu YC, Wang Z, Pei ZF, Davies M (2016) Measured energy use and indoor environment quality in green office buildings in China. Energy Build. 129:9–18
5. American National Standards Institute (ANSI)/American Society of Heating, Refrigerating and Air-Conditioning Engineers (ASHRAE), Standard 55-2013 Thermal Environmental Conditions for Human Occupancy, Atlanta: ASHRAE, 2013
6. Ministry of Housing Urban and Rural Development (2012) Design Code for Heating, Ventilation and Air Conditioning in Civil Buildings (GB50736-2012). China Architecture & Building Press, Beijing, China (In Chinese)
7. Ministry of Ecology and Environment, Indoor air quality standard (GB/T 18883-2002), Beijing, China, 2002 (In Chinese)
8. Architecture Society of China, Assessment standard for healthy building (T/ASC02-2016), Beijing, China, 2016 (In Chinese)
9. Ministry of Housing Urban and Rural Development (2013) Standard for Lighting Design of Buildings (GB 50034-2013). China Architecture & Building Press, Beijing, China (In Chinese)

Chapter 5
Elementary Operating Performance Diagnosis of Individual Green Office Building Based on Poor-Quality Data

5.1 Background

Starting from this chapter, the research content will shift from the macroscopic characteristics to the microscopic characteristics of the building operating performance, focusing on the performance analysis and diagnosis of individual building cases. When carrying out case studies, the data quality and requirement often determine the depth of our research. This chapter mainly focuses on the basic analysis method with low-quality building performance data.

In the previous literature, low-quality data is often a big challenge for the further diagnosis of building performances. It is difficult to conduct an effective data mining work with poor data. In order to solve this problem, this chapter will propose several statistical regression models for the building performance diagnosis with low-quality data. The proposed models will have a clear physical meaning and strong universality, which can help dig out the hidden value from the low-quality data and extract useful information for further building performance diagnosis and optimization.

The research content is divided into two parts: (1) The building energy diagnosis model with monthly energy bills and weather data; (2) The IEQ diagnosis model with subjective survey data.

5.2 Building Energy Diagnosis Through the IMT Model

5.2.1 Introduction of the IMT Model

The proposed building energy performance diagnosis method is inspired by ASHRAE Inverse Modeling Toolkit (IMT) [1]. The previous studies [2–9] mainly used IMT to solve two types of problems. The first one was to calibrate energy use calculation by weather conditions. For example, in the study of Kissock and Eger

Y. Geng, *Data-Driven Research on Characteristics of Operating Performance of Green Office Buildings*, Springer Theses, https://doi.org/10.1007/978-981-95-2329-0_5

[7], IMT tool was mainly used to calculate the weather-adjusted energy savings between the pre- and post-retrofit periods. The second one was to analyze building energy performance efficiently. For example, Reichmuth et al. [8] and Abels et al. [9] applied IMT to disaggregate total energy use into several parts and identify other building attributes, which can be compared with reference value to find major building inefficiencies.

However, in this study, the IMT tool is developed to diagnose the performance gap between actual energy use and designed (or simulated) energy use, and to find out the reasons causing the difference. Thus, we apply IMT tool to analyze not only the actual energy data but also the simulated energy data, and then compare them to see the difference. Based on the building attributes derived from IMT tool, we can locate the key reasons and parameters causing the difference. Furthermore, detailed measurements are also conducted to test the diagnosis results and to find out how much the actual parameter changes compared to the design value, which can also feedback to the simulation model for calibration. This process of validation makes the diagnosis more convincing, but it was rarely mentioned in the previous studies. This new application of IMT tool makes it easier.

The IMT method is involved with regression models between energy use and outside weather. In general, the relationship between building energy consumption and outdoor air temperature throughout the whole year can be classified into six models, as shown in Table 5.1.

The above regression models not only show the relationship between building energy use and outside weather, but also reveal other important information about building characteristics and the operation of cooling/heating systems. To take 5P Model as an example, this model describes that building energy use keeps the constant baseload level (i.e., b_0) until the cooling or heating change-point temperature (i.e., c_1 or c_2) is reached. When outdoor air temperature is higher or lower than the cooling or heating change-point temperature, building energy use increases linearly at the cooling or heating slope (i.e., a_1 or a_2). These five coefficients of the model are directly influenced by building physical properties and the operation and efficiency of building systems, thus they can be used for further diagnosis of building energy performance. The coefficient b_0 indicates the weather-independent energy consumption throughout the year, which is mainly consist of the energy use of lighting, plug and other auxiliary systems. The coefficient a_1 and a_2 indicate energy use associated with outdoor weather, which is mainly consumed by cooling and heating systems. The absolute value of these two coefficients is affected by cooling/heating loads and the efficiency of cooling/heating system. The coefficient c_1 and c_2 reflect the change-point outdoor air temperature when cooling or heating system begins to operate. The operation schedule and set-point temperature of cooling or heating system, as well as other information about internal loads and envelope properties of the building, can be inferred from the coefficient c_1 and c_2.

Therefore, this study applies IMT to conduct regression analysis between building energy use and weather data and uses the key coefficients to identify energy performance at the system level. Another strength of this building energy diagnosis method

Table 5.1 Regression models between building energy consumption and outdoor air temperature

Model type	Schematic plot	Description	Meaning and example
1P model	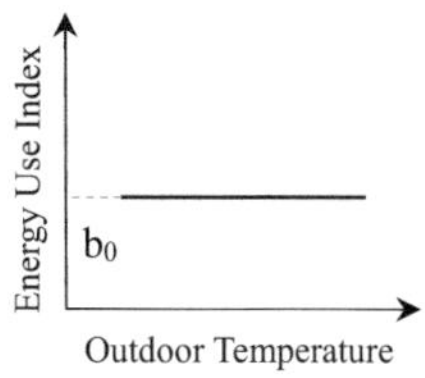	Building energy use has little to do with the outdoor climate	(1) Buildings whose HVAC system consumes a small proportion of total energy consumption (2) Special building type, such as data center
2P model	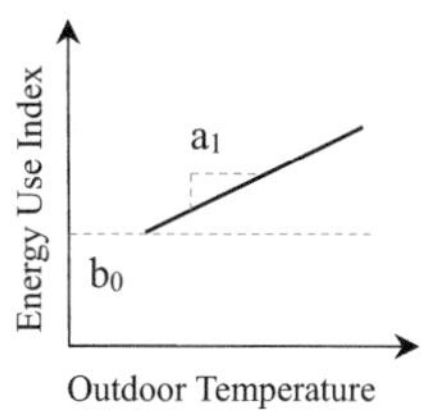	Building energy use is positively correlated with outdoor air temperature	(1) This model means the cooling system runs throughout the year (2) Some buildings in hot-summer warm-winter zone in China
3P model	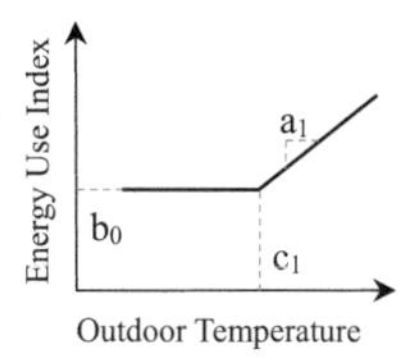 (3PC) Or 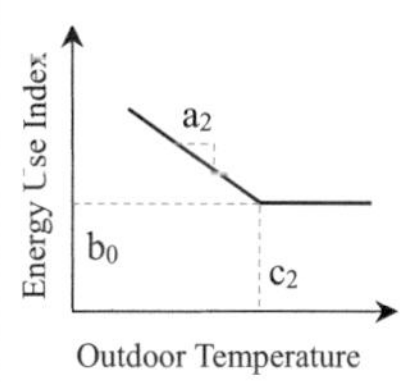 (3PH)	Building energy use rises when outdoor air temperature is higher (or lower) than a certain value	(1) When outdoor air temperature is higher (or lower) than a certain value, the cooling (or heating) system will be on operation (2) 3PC: Buildings with cooling demand in summer and without heating demand in winter (3) 3PH: Buildings with heating demand in winter and without cooling demand in summer

(continued)

Table 5.1 (continued)

Model type	Schematic plot	Description	Meaning and example
4P model	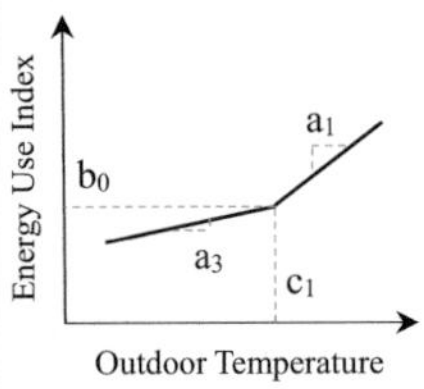	Building energy use rises with outdoor air temperature at a smaller slope. When temperature exceeds a certain value, building energy rises at a larger slope	(1) The larger slope corresponds to the period of active cooling, while the smaller slope corresponds to the period of passive cooling with air side economizer (2) VAV system with air-side economizer and inner zone load is significant, which means the building also has cooling demand in winter and transition seasons
5P model	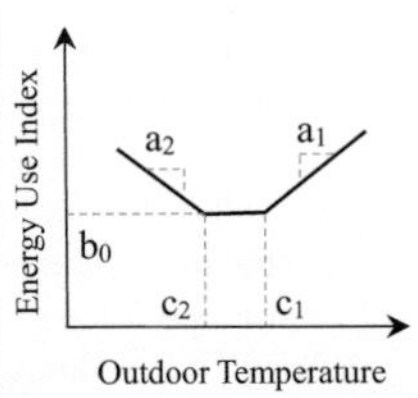	In a certain range of outdoor air temperature, building energy use is relatively constant. When the temperature is below or exceeds the range, building energy use will increase	(1) Cooling system operates in summer, heating system operates in winter, while HVAC system is off in transition seasons (2) Most buildings in cold zone or hot-summer cold-winter zone in China

is that it does not require detailed energy data. All required information is the monthly energy bills and weather data, which can be easily obtained.

5.2.2 Diagnosis Method and Framework

The overall outline of the building energy performance diagnosis is shown in Fig. 5.1. The diagnosis procedure consists of two parts. The first part is the regression analysis between energy use and outside weather, which can obtain the key coefficients reflecting further information about the building systems. The second part is energy performance evaluation and benchmarking. There are generally two methods of benchmarking: (1) to compare with expected performance in the design stage, which is simulated by the reference model [10]. Based on the whole diagnosis procedure, poor energy performances can be located and the key points of energy saving will be targeted to improve the performance.

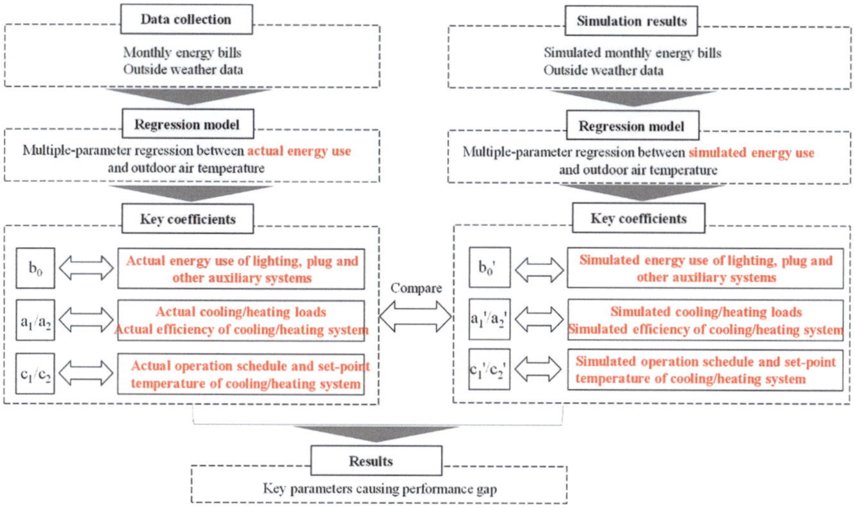

Fig. 5.1 The outline of building energy performance diagnosis

5.2.3 Case Study

In order to test the application of the energy performance diagnosis method based on building energy bills and weather data, one case study of an office building is conducted. The office building is in Tianjin, China and it belongs to cold climate zone. Figure 5.2 displays the weather condition of Tianjin in the typical meteorological year. This building was renovated in 2012 and it gained the 3-star certification of green building after refurbishment. The total floor area of the building is 5700 m^2 and the height is 25 m. Other basic information of the case study building can be found in Table 5.2.

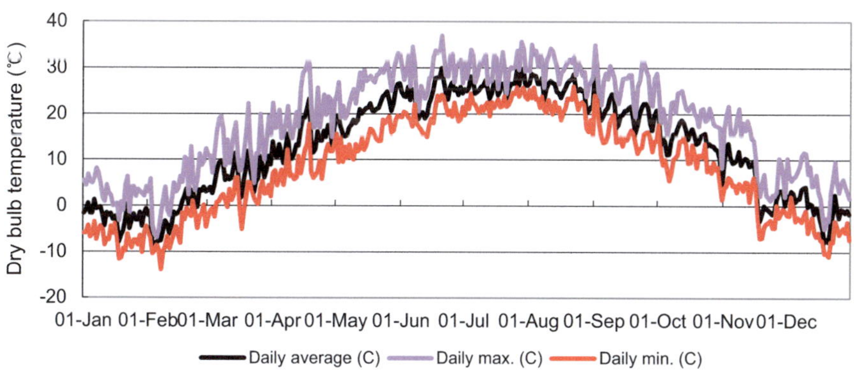

Fig. 5.2 Daily outdoor air temperature of the typical meteorological year of Tianjin

Table 5.2 General information of case study buildings

Basic information	
Photos	
Building type	Office
Location/climate zone	Tianjin, China/cold zone
HDD/CDD	HDD18 > 2735, CDD26 > 56
Floor area (m^2)	5700
Height (m)/floor	25/6
Occupant number	300
Envelope	
WWR	0.4 (South), 0.2 (North, West and East)
U-value (W/m^2 K)	2.3 (Window), 1.1 (Curtain Wall)
HVAC system	
Cooling/heating source	Ground source heat pump + Water storage system
Terminals	Radiant floor + Fan coil units
Ventilation	Mix-mode (mechanical + natural ventilation)

In order to conduct building energy performance diagnosis based on the regression model, annual monthly energy bills during 2016 were collected from the building owner, as shown in Table 5.3. Besides that, this office building is also an information rich building with energy consumption monitoring system. The sub-metering technology is applied to record energy use data of each individual system, as well as main components. Some important operating parameters of HVAC system are also monitored in real time, such as temperature and flow rate of supply and return water, part load ratio and energy efficiency ratio of the cooling or heating system, operating status of fans/ pumps, etc. Although the advanced energy consumption monitoring system provides detailed energy use and performance data at the system and component level, few of them are required the diagnosis procedure. Instead, the main value of these monitoring data is to test the diagnosis results based on monthly energy bills and weather data, which will be presented in Sect. 5.2.4.

Based on the monthly building energy bills and weather data, the multiple-parameter regression can be conducted to obtain the key coefficients reflecting building energy performance. Figure 5.3 and Table 5.4 show the regression results between monthly average energy use and outdoor air temperature. It can be found that the relationship is highly consistent with 5P model ($R^2 = 0.94$), which illustrates that

Table 5.3 The input parameters of reference model for benchmarking

Month	Monthly energy bills (kWh)	Num. of workdays	Average daily energy use (kWh/day)
Jan	41,865	20	2093
Feb	27,025	16	1689
Mar	27,339	23	1189
Apr	16,823	20	841
May	17,238	21	821
Jun	20,968	20	1048
Jul	30,274	21	1442
Aug	36,295	23	1578
Sep	14,932	20	747
Oct	14,959	17	880
Nov	29,835	21	1421
Dec	38,753	23	1685

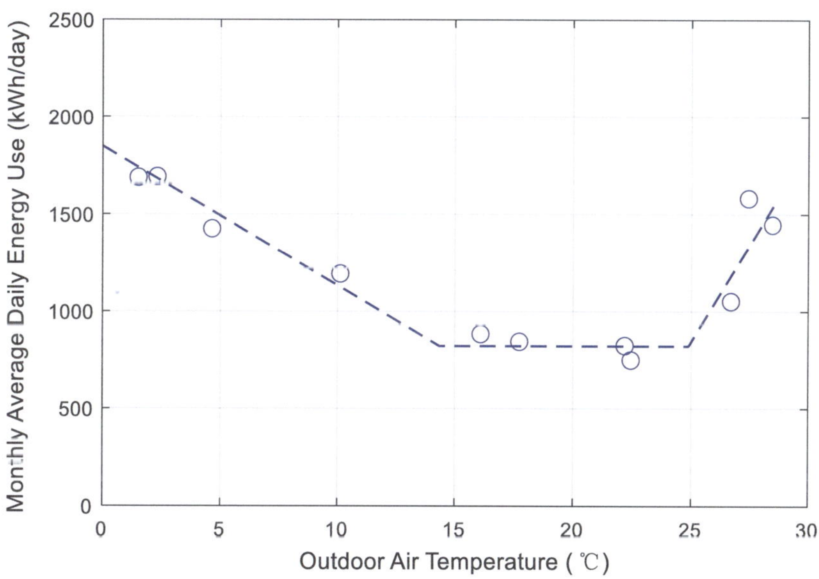

Fig. 5.3 The regression results between monthly average energy use and outdoor air temperature

building energy use is relatively constant in the middle range of outdoor air temperature, while energy use increases as outdoor air temperature goes beyond the range. Basically, this relationship also reveals the operation mode of HVAC system in the whole year, i.e. cooling system operates in summer and heating system operates in winter, while HVAC system remains closed in transition seasons.

Table 5.4 The regression coefficients and goodness of fit

Model	$E = \begin{cases} b_0 + a_1(T - c_1) & T > c_1 \\ b_0 & c_2 < T \le c_1 \\ b_0 + a_2(c_2 - T) & T \le c_2 \end{cases}$ (5P Model)	
Regression coefficients	b_0 (Baseload)	822.2
	a_1 (Cooling slope)	199.9
	a_2 (Heating slope)	71.6
	c_1 (Cooling change-point)	24.9
	c_2 (Heating change-point)	14.3
Goodness of fit	R^2	0.94

According to the regression coefficients, some initial results of energy performance diagnosis can be reported. Firstly, the baseload b_0 represents energy use of lighting, plug and other auxiliary systems and it can be used to break down the annual and monthly building energy consumption into two main systems, as described in Eqs. (5.1) and (5.2).

$$E_{HVAC,m} = E_m - E_{Others,m}, \quad E_{Others,m} = b_0 \cdot d_m \tag{5.1}$$

$$E_{HVAC,a} = E_a - E_{Others,a}, \quad E_a = \sum E_m, \quad E_{Others,a} = b_0 \cdot \sum d_m \tag{5.2}$$

where, E_a/E_m is the annual/monthly building energy consumption; $E_{HVAC,a}/E_{HVAC,m}$ is the annual /monthly energy consumption of HVAC system; $E_{Others,a}/E_{Others,m}$ is the annual/monthly energy consumption of other systems, including lighting, plug and auxiliary systems; d_m is the number of workdays in each month; and b_0 is the baseload coefficient.

Figure 5.4 shows the break-down results of annual and monthly building energy use. In the months of April, May, September and October, there is almost no energy consumption of HVAC system. However, in the months of January, December and August, the HVAC system consumes lots of energy, which counts for 61, 51 and 48% of the total energy consumption in the corresponding month. Throughout the whole year, lighting, plug and auxiliary systems take up 63.4% of the annual building energy consumption and HVAC system consumes the remaining 36.6% energy.

Secondly, the cooling/heating change-point c_1 and c_2 reflect when cooling or heating system begins to operate. According to the regression results of c_1 and c_2, HVAC system does not operate when the outdoor air temperature is within the range of 14.3–24.9 °C, cooling system begins to operate when the outdoor air temperature rises to about 24.9 °C and heating system begins to operate when the outdoor air temperature decreases to about 14.3 °C. After the comparison with weather data in the whole year, it can be approximately inferred that the operation of cooling system

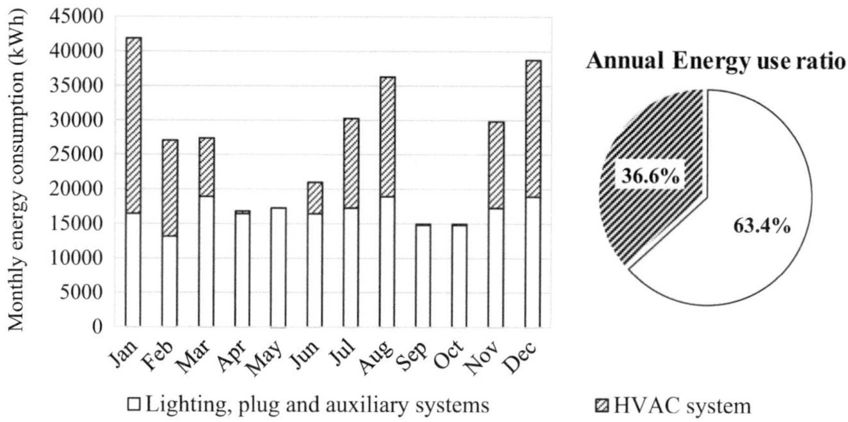

Fig. 5.4 The division of annual and monthly building energy consumption

in a year lasts from the beginning of June to the end of August, and the heating system operates from the beginning of November to the end of March.

Thirdly, the cooling/heating slope a_1 and a_2 can be used as the indicators of cooling/heating system performance. However, the coefficient itself can provide little useful information if there is no benchmark data for comparison. In order to examine whether the actual building energy performance reaches the expected goal, energy simulation result at the design stage is selected as the benchmark data to conduct further diagnosis in the following studies. The simulation is based on the reference model and the detailed information about simulated reference model is shown in Appendix C. After building energy simulation has been done, the monthly simulated energy consumption is used as the input of multiple-parameter regression model with the same procedure of regression. Finally, the regression result of the simulated energy consumption serves as benchmarks for actual energy performance and is compared to the regression result based on the energy bills. The benchmarking process can further diagnose the performance gap between the operation stage and the design stage and help locate the poor performance and key points of energy-saving.

Figure 5.5 shows the comparison of the regression results of actual energy data and simulated energy data. As can be seen from the figure, there are three main differences between the two regression results. Firstly, the actual baseload (822 kWh/day) is lower than the simulated baseload (934 kWh/day), which indicates that the lighting, plug and auxiliary systems save energy than the design. One possible reason is that the designed power density of lights and equipment tends to be a large value in order to ensure the adequate capacity of HVAC system, while the power density is lower in real conditions. Another reason might be concerned with the schedule of lighting, plug and auxiliary systems, which means the total running hours in a year is less than design. The second difference is the cooling change-point. In the designed condition, cooling system operates when outdoor air temperature is higher than 20.7 °C (the corresponding period is from the middle of May to the

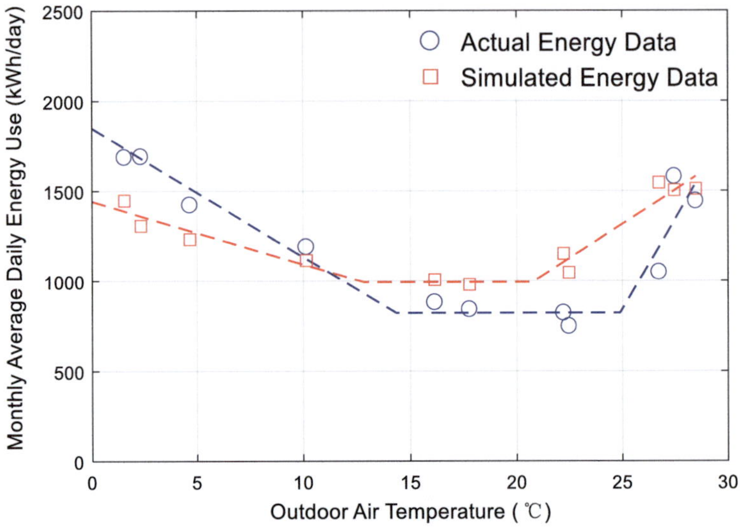

Fig. 5.5 The comparison of the regression results of actual energy data and simulated energy data (benchmark data)

end of September), whereas the operation of cooling system is actually from the beginning of June to the end of August. This difference is mainly caused by the mix ventilation mode in actual conditions. During transition seasons and early or late summer, natural ventilation is used to improve the indoor thermal environment, instead of active air conditioning. Thus, the cooling system can be closed when it is not very hot outside, which also saves energy consumption. The third deviation is reflected in the cooling/heating slope and the regression result of actual energy data shows relatively bigger slope on both cooling and heating sides. The slope is related to cooling or heating loads, plus cooling or heating system efficiency. Because envelop parameters and weather data under the actual and simulated conditions are basically the same, the influence of external load can be roughly eliminated. In terms of the influence of internal load, the past experience and knowledge demonstrate that larger internal load would lead to bigger slope in cooling seasons and smaller slope in heating seasons. However, the cooling and heating slopes of actual energy data are consistently bigger than those of simulated energy data, which illustrates internal load is not the main cause of the difference of cooling/heating slope. Thus, the slope is primarily affected by cooling or heating system efficiency and the steeper slope means more energy consumption with weather variations, i.e. lower efficiency of cooling or heating system. The comparison of slope indicates that the efficiency of cooling/heating system is not as good as design. The above content is the energy performance diagnosis results at the system level and these diagnosis conclusions and inferred reasons will be further validated in Sect. 5.2.4.

5.2.4 Validation

In this section, the performance data from advanced energy consumption monitoring system together with field surveys and measurements are used to test the diagnosis results based on the regression model between monthly energy bills and weather data. The validation results are shown as follows.

Figure 5.6 presents the comparison between the measured and calculated energy use break-down results. The calculated data based on the diagnosis model is well consistent with the measured data from the sub-metering systems. The relative deviations of energy consumption break-down results in each month are all controlled within the range of 15% and the annual average relative error is 2.5%.

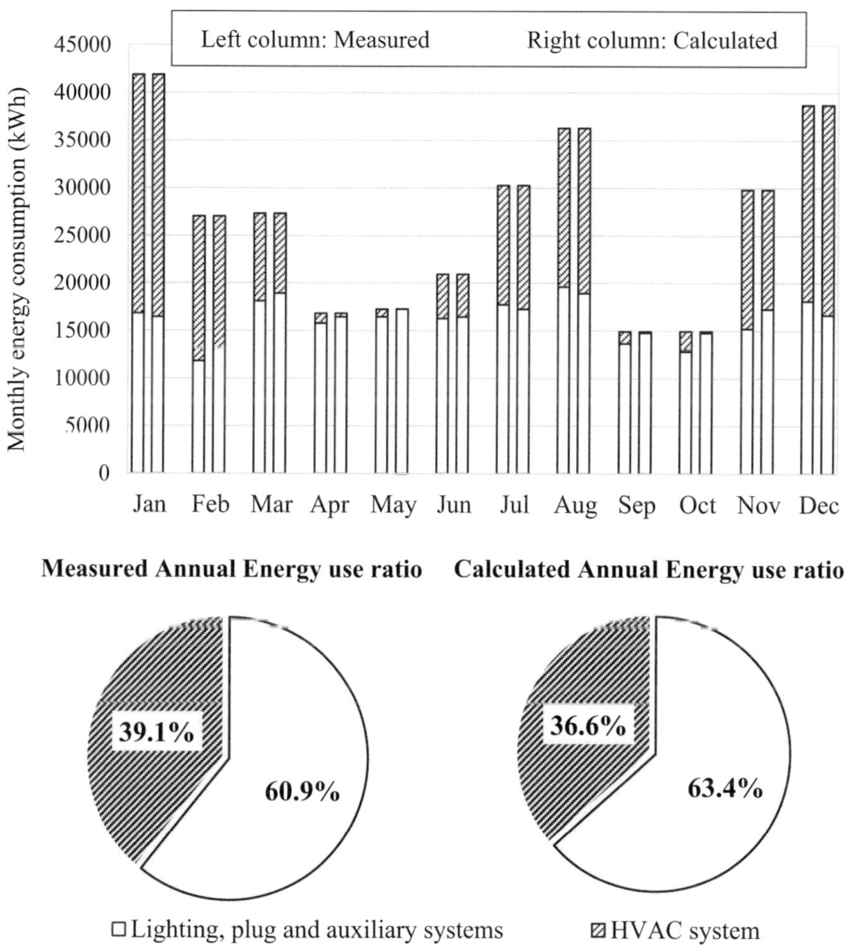

Fig. 5.6 The comparison between the measured and calculated energy use break-down results

The diagnosis result by the baseload coefficient also indicates that lighting, plug and auxiliary systems actually consume less energy than the design. In order to valid this conclusion, the power density and schedule of lighting, plug and auxiliary systems in real conditions are investigated and compared to the designed data. The designed lighting and equipment power density are 9 W/m^2 and 13 W/m^2, however the investigated power density of lighting system and equipment are quite smaller with actual value of 7.5 W/m^2 and 10 W/m^2 respectively. In addition, the actual schedules of lighting system and equipment are different from the reference mode of simulation, as shown in Fig. 5.7. The actual schedule was investigated through a typical week and the average data of weekdays is presented and compared to the reference schedule. It can be observed that lights and equipment are actually used less frequently on daytime. For example, half of offices don't need to turn on the lights until 3:00 pm and only 80% equipment is on the running state during daytime. Whereas overtime work tends to be common in this building, so lights and equipment are often used at night. In general, the overall use time of lights and equipment during the whole day are fewer in real conditions (Lighting: 9.3 h vs. 11.2 h; Equipment: 9.4 h vs. 9.4 h). Therefore, the surveyed data proves that lighting and plug systems save energy than the design.

The diagnosis result by cooling/heating change-point illustrates that the operation of cooling system in a year lasts from the beginning of June to the end of August, and the heating system operates from the beginning of November to the end of March. Furthermore, the benchmarking process shows that the operation period of cooling system is relatively shorter than the design and the possible reason is inferred to be the mix ventilation mode during transition seasons and early or late summer. Table 5.5 shows the actual operating timetable of cooling and heating system in the whole year and it is basically consistent with the analysis result by the diagnosis model. Furthermore, in order to test whether natural ventilation is applied to improve the indoor thermal environment in transition seasons and early or late summer, long-term instrumental measurement of indoor CO_2 concentration from April to October is conducted, as shown in Fig. 5.8. It can be found that the average level of indoor CO_2 concentration in April, May, September, and October is significantly lower than that in June, July and August. The difference passes *t*-test with the significant level of $p < 0.05$. Thus, the measured CO_2 concentration verifies that natural ventilation is commonly used during transition seasons, which agrees with the previous inference by diagnosis model.

In order to valid the diagnosis result by cooling/heating slope that the actual efficiency of cooling/heating system is not as good as the design, measurements of HVAC system and main components are conducted during the cooling season in 2016. Figure 5.9 presents the measured performance of ground source heat pump (GSHP) in a typical week. As can be seen from the figure, both part load ratio (PLR) and COP of GSHP vary greatly during the measurement. The measured result shows that PLR varies from 9% to nearly 100% and COP varies from 1.15 to 7.03. For most of time, GSHP operates under the condition of PLR below 60% and the efficiency is significantly lower than the rated value. On average, the mean value of PLR and COP during the measured period is 51% and 4.51 respectively.

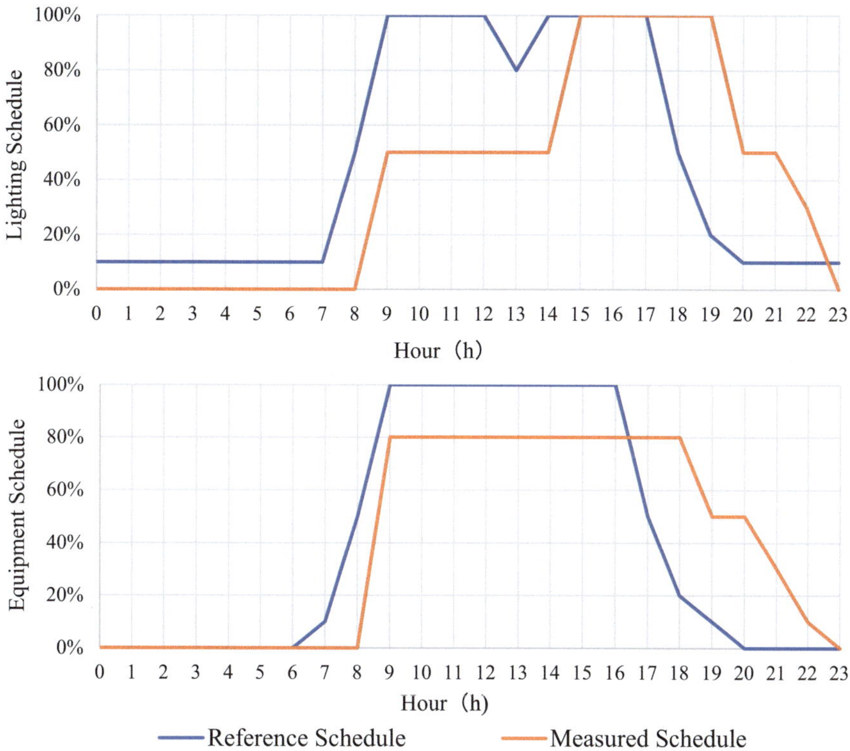

Fig. 5.7 The comparison between the reference schedule and measured schedule of lighting system and equipment

Furthermore, the performances of HVAC system and sub-systems in the whole cooling season are calculated based on the data from the advanced monitoring system, as illustrated in Table 5.6. Except for the EER of GSHP and cooling water system, the efficiency of other sub-systems is lower than benchmark data. The overall efficiency of the whole HVAC system during the cooling season is 3.58, which is also slightly lower than the reference value in the simulation model. Above all, the detailed analysis proves that the actual performance of cooling/heating system does not reach the expectation due to the low PLR in most of time.

5.2.5 Conclusion

This section presents a simplified but accurate method for building energy performance diagnosis which can assess the energy performance from building level to system level with only limited information. Based on multiple-parameter regression between the whole building energy bills and outdoor weather data, the proposed

Table 5.5 The actual operating timetable of cooling and heating system in the whole year

Season	Period	Days	Strategy	Detailed method
Summer	Jun. 10th–Aug. 31th	83	GSHP A + B	GSHP A produces 7 °C low-temperature cold water for dehumidification GSHP B produces high-temperature cold water for cooling
Winter	Oct. 21st–Nov. 15th	26	GSHP B	GSHP B produces hot water to the floor radiant heating system
	Nov. 16th–Mar. 31st	137	GSHP A + B	Daytime: GSHP B produces hot water to the floor radiant heating system Nighttime: GSHP A produces energy which is stored in the energy storage system and the stored energy will release during daytime
Transition seasons	Sep. 1st–Oct. 20th and Apr. 1st–Jun. 9th	120	NV	Windows are opened to expel internal heat
			NV + Fans	NV and mechanical convection are combined
			Free cooling	Water in GSHP buried pipe is pumped to the floor radiant system for free cooling

Note: *GSHP* presents *Ground Source Heat Pump* and *NV* presents *Natural Ventilation*

diagnosis method uses the regression coefficients to identify the energy use of main systems (i.e., lighting-plug system and cooling/heating system), as well as other detailed information about physical properties of the building and cooling/heating system (i.e., cooling or heating load, the operation of cooling/heating system and the efficiency of cooling/heating system).

To be specific, the baseload coefficient indicates the weather-independent energy consumption throughout the year, which is mainly consist of the energy use of lighting, plug and other auxiliary systems. The cooling/heating slopes indicate energy use associated with outdoor weather, whose values are mainly affected by cooling/heating loads and the efficiency of cooling/heating system. The cooling/heating change-points reflect the outdoor air temperature when cooling or heating system begins to operate and their values are influenced by the operation schedule and set-point temperature of cooling or heating system, as well as other information about internal loads and envelope properties of the building.

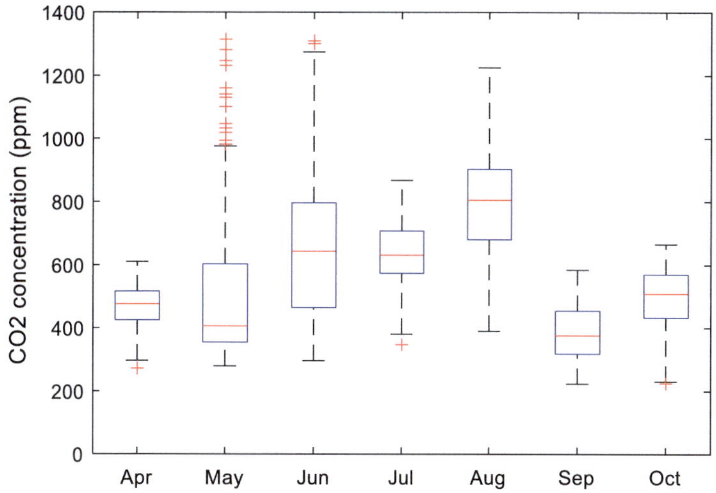

Fig. 5.8 The distribution of indoor CO_2 concentration in transition seasons and summer

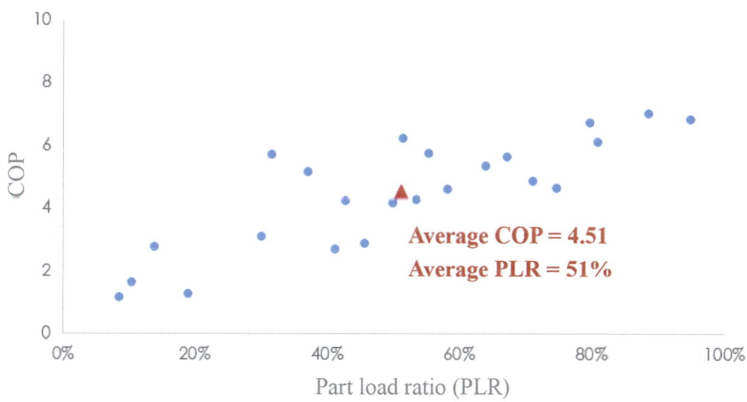

Fig. 5.9 The performance of GSHP in a typical week during the cooling season

Table 5.6 The performances of HVAC system and sub-systems in the whole cooling season

Indicator	Performance in the whole cooling season	Benchmark
EER of the whole HVAC system	3.58	–
EER of chiller plant	2.80	≥ 3.8
EER of GSHP	4.79	≥ 4.2
EER of chilled water system	29.86	≥ 30
EER of cooling water system	58.07	≥ 25
EER of terminals	18.87	≥ 24 (FCU)

One case study was conducted in an office building in China to test the application of this diagnosis method. In addition, the regression result of simulated energy consumption served as benchmark data to further diagnose the performance gap between the operation stage and the design stage. Finally, all diagnosis results have been verified by the performance data from advanced energy consumption monitoring system together with field surveys and measurements.

Although the proposed building energy performance diagnosis method can help locate the poor performance and key points of energy-saving with limited energy data, more detailed energy performance data at the system and subsystem/component level needs to be measured and analyzed to get more quantitative diagnosis results. Besides that, the accuracy of the diagnosis method and its application to more types of buildings are important issues for future research.

5.3 IEQ Diagnosis Through the Kano Model

5.3.1 Introduction of the Kano Model

Kano's model of satisfaction [11] belonging to marketing research was brought into IEQ domain by Kim and de Dear [12, 13] and was tested suitable in the context of building occupants' IEQ satisfaction. The aim of adapting the Kano's model to the building context is to better understand how various IEQ factors influence on occupants' overall satisfaction in different ways. Then three types of relationship between individual IEQ factors and overall satisfaction were defined and categorized, as shown in Fig. 5.10.

(1) Basic Factors: Excellent performance of them doesn't necessarily bring about increment in overall satisfaction but they generate overall dissatisfaction if not fulfilled.
(2) Bonus Factors: Inferior quality of these factors has little negative influence on occupants' dissatisfaction. However, they have a significantly positive effect on overall satisfaction when they are well-performed.
(3) Proportional Factors: The changes of these factors have a proportional effect on the level of occupants' overall satisfaction.

Through the analysis of occupant survey samples from CBE's database [14], 15 IEQ factors have been labelled as Basic Factors (including 'temperature', 'noise level', 'amount of space', 'visual privacy', 'adjustability of furniture', 'colors & textures' and 'workspace cleanliness') and Proportional Factors (including 'air quality', 'amount of light', 'visual comfort', 'sound privacy', 'ease of interaction', 'comfort of furnishing', 'building cleanliness' and 'building maintenance') [12]. It was also found that occupants' expectations and responses to IEQ performance can differ between buildings with different ventilation types (i.e. Air-Conditioned,

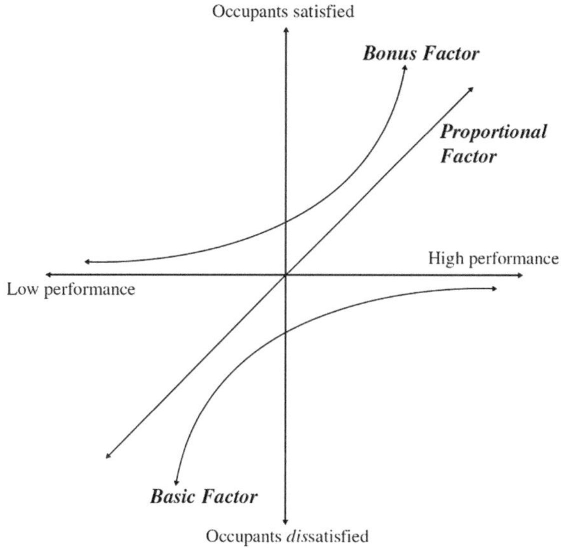

Fig. 5.10 Kano model diagram

Mixed-Mode and Naturally Ventilated), thus the classification of some IEQ factors into Basic, Bonus and Proportional groups will change [13].

These researches are innovative, with a total of 43,021 [12] and 22,518 [13] survey samples. However, all data are from the foreign database. Whether the findings are appropriate for the Chinese green office buildings is still unknown. Therefore, this study aims to find out the effect of different IEQ factors on overall satisfaction in Chinese green office buildings, based on the Kano model.

5.3.2 Modeling Methodology

Firstly, questionnaire samples are divided into three groups in terms of each IEQ factor's performance level; (a) ***satisfied group***: occupants who are distinctly satisfied with the IEQ factor (those who rate their satisfaction level with + 3 or + 2), (b) ***dissatisfied group***: occupants who are distinctly dissatisfied with the IEQ factor (those who rate their satisfaction level with − 3 or − 2), and (c) ***reference group***: occupants who remain neutral on the IEQ factor performance (those who rate their satisfaction level with − 1, 0, or + 1). The logic of this classification refers to that used by Fanger [15] in his mapping from 7-point scale of thermal sensation (PMV) to thermal satisfaction/dissatisfaction bifurcation (forming the basis of PPD index). Based on this sorting, the difference of overall satisfaction between these three groups can be further studied.

Secondly, two dummy variables with a binary coding (0 and 1) are applied to represent three groups of each IEQ factor (i.e. *satisfied*, *dissatisfied* and *reference group*), as Eq. 5.3. One number pair of 1, 0 is assigned to the '*satisfied group*', another number pair of 0, 1 is assigned to '*dissatisfied group*' and the last number pair of 0, 0 is assigned to '*reference group*'. This process is repeated across all 6 IEQ factors.

$$X_{1,factor\,N} = \begin{cases} 1, & satisfied \\ 0, & otherwise \end{cases} \quad X_{2,factor\,N} = \begin{cases} 1, & dissatisfied \\ 0, & otherwise \end{cases} \tag{5.3}$$

Thirdly, the individual IEQ factors with dummy variables are defined as independent variables, the 'overall satisfaction' is defined as dependent variable, and multiple regression analysis is conducted as Eq. 5.4.

As shown in the regression Eq. 5.4, two coefficients for each IEQ factor have been created: a_1 to measure the IEQ factor N's impact on overall satisfaction when the performance level of IEQ factor N is high (*positive impact*), and a_2 to measure the IEQ factor N's impact on overall satisfaction when the IEQ factor N performed poorly (*negative impact*). By comparing the absolute values of these two regression coefficients, one particular IEQ item can be identified *Basic*, *Bonus* or *Proportional Factor*, according to Kano's model. For example, if the absolute value of a_2 for IEQ factor N outweighs that of a_1 (*negative impact > positive impact*), IEQ factor N has a stronger impact on occupants' overall satisfaction when the performance of IEQ factor N is rated as poor, then this IEQ is identified as *Basic Factor*. Conversely, if the absolute value of a_1 for IEQ factor N outweighs that of a_2 (*positive impact > negative impact*), IEQ factor N is categorized into *Bonus Factor*. Finally, if the absolute values of a_1 and a_2 for IEQ factor N are approximately equal (*positive impact = negative impact*), IEQ factor N is defined as *Proportional Factor*.

$$Y = a_0 + \sum_{N=1}^{6} \left(a_{1,factorN} \cdot X_{1,factorN} + a_{2,factorN} \cdot X_{2,factorN} \right) \tag{5.4}$$

Y: Occupants' overall satisfaction level with the whole indoor environment.

a_0: Average of overall satisfaction score of reference groups.

$X_{1,\text{factor N}}$: Dummy variable for satisfied group of IEQ factor N.

$X_{2,\text{factor N}}$: Dummy variable for dissatisfied group of IEQ factor N.

$a_{1,\text{factor N}}$: Regression coefficient for satisfied group of IEQ factor N.

$a_{2,\text{factor N}}$: Regression coefficient for dissatisfied group of IEQ factor N.

5.3.3 Case Study and Results

The author selected a green office building in the HSCW zone for the case study. The IEQ satisfactions of private office and open-plan office were investigated respectively. Based on the Kano model, demands and preferences of the two occupant groups on various IEQ factors were revealed, which clarify the key factors for the overall IEQ improvement.

According to the method in Sect. 5.3.2, the author established a multiple regression model based on the survey data, and finally obtained the key parameters a_1 and a_2 that can reflect the quantitative relationship between IEQ factors and overall satisfaction. Table 5.7 shows the regression results. The value of a_1 reflects the impact of a certain IEQ factor on overall satisfaction when it performs well, and the value of a_2 reflects the impact of a certain IEQ factor on overall satisfaction when it performs poor.

Furthermore, the author ranks the IEQ factors according to their impacts on the overall satisfaction (i.e., the absolute value of a_1 or a_2), as shown in Fig. 5.11. The rankings correspond to two scenarios when the performance of a certain IEQ factor is good or poor.

It can be found that:

(1) For occupants in both private and open-plan offices, acoustics and space have the greatest impacts on overall satisfaction, especially for the negative impact on overall satisfaction when the acoustic environment performs poorly and the positive impact on overall satisfaction when the space quality is good (both ranking No.1).
(2) Poor thermal performance has a relatively strong negative impact on the overall satisfaction in private offices, while good thermal performance has a relatively strong positive impact on overall satisfaction in open-plan offices.
(3) The impact of IAQ on overall satisfaction in open-plan offices is greater than that in private offices, no matter IAQ performs well or poorly.

By comparing the absolute values of a_1 and a_2, we can divide various IEQ factors into Basic, Bonus or Proportional factor, as shown in Table 5.8.

For private office occupants, due to their high expectations for the indoor environment, the categories of each IEQ factor are all Basic or Proportional. Thermal comfort, lighting and acoustics are Basic factors, indicating that these IEQ factors have a greater negative impact on overall satisfaction when they perform poorly, while the positive impact on overall satisfaction is relatively small once they meet

Table 5.7 The multiple regression results based on the Kano model

IEQ factor	Private office			Open-plan office		
	a_1	a_2	Difference	a_1	a_2	Difference
Thermal comfort	0.06*	– 0.25*	Significant	0.16*	– 0.18*	Not significant
IAQ	0.18*	– 0.22*	Not significant	0.25*	– 0.25*	Not significant
Lighting	0.16*	– 0.22*	Significant	0.09**	– 0.20**	Significant
Acoustics	0.21*	– 0.38*	Significant	0.15*	– 0.28**	Significant
Space	0.32*	– 0.29*	Not significant	0.44*	– 0.24*	Significant
Service	0.10	– 0.11	Not significant	0.03	– 0.05	Not significant

*$P < 0.05$, **$P < 0.01$

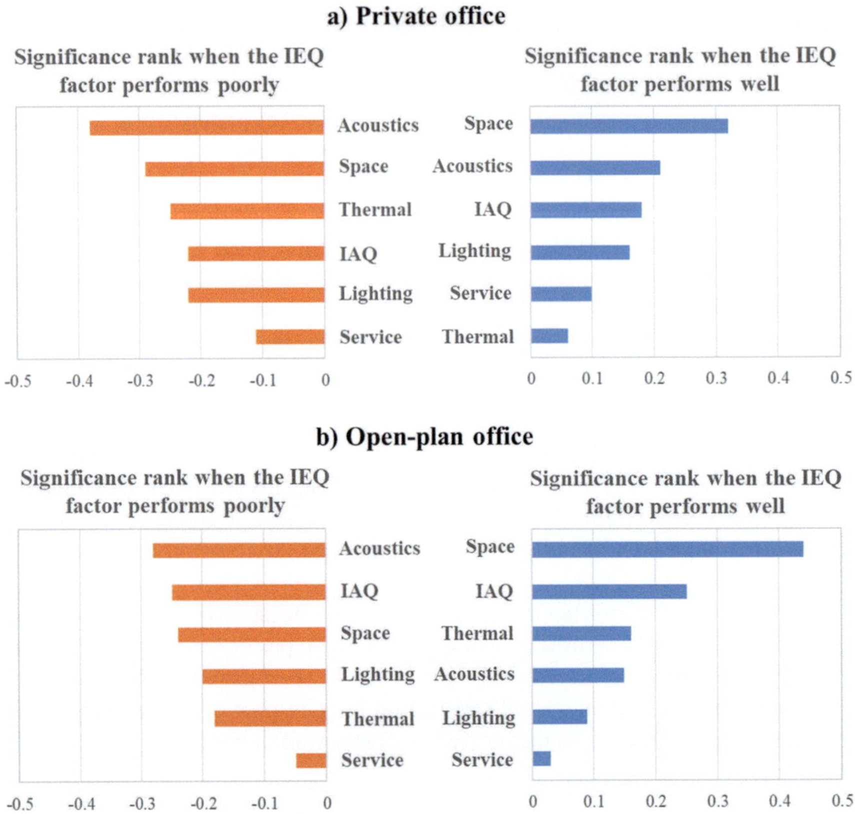

Fig. 5.11 The IEQ factor ranking according to the impact on overall satisfaction

Table 5.8 The classification results of IEQ factors for private and open-plan office occupants

IEQ factor	Private office	Open-plan office
Thermal comfort	Basic	Proportional
IAQ	Proportional	Proportional
Lighting	Basic	Basic
Acoustics	Basic	Basic
Space	Proportional	Bonus
Service	Proportional	Proportional

the basic requirements. Therefore, the most efficient way for the control of these IEQ factors is to let them just meet the criteria, and any effort beyond that is unnecessary. Air quality, Space and building service are Proportional factors, indicating that there is a linear relationship between their performance and the overall satisfaction. Therefore, the more effort on the improvement of these IEQ factors, the higher overall satisfaction can be obtained.

For open-plan office occupants, the relationship between some IEQ factors (IAQ, lighting, acoustics, and building service) and the overall satisfaction are the same as above. However, other IEQ factors show different characteristics due to the changes of occupant expectations and requirements. For example, thermal comfort changes from Basic factor in private offices to Proportional factor in open-plan offices, and space changes from Proportional factor in private offices to Bonus factor in open-plan offices. These differences indicate that the thermal environment and space condition in open-plan offices may not be as good as those in private offices. Therefore, open-plan office occupants have lower expectations for these IEQ factors, compared with private office occupants. In other words, open-plan office occupants tend to be more tolerant when the thermal or space condition is poor, and feel more satisfied when the thermal or space condition becomes superior.

5.4 Summary

This chapter proposes two regression models for the low-quality building performance data which are difficult to be deeply analyzed in previous studies. The proposed models have a clear physical meaning and strong universality. They can be used to identify the typical building performance problems with the low-quality data, providing a scientific reference for the further performance optimization.

- Building energy diagnosis model

The proposed model used building energy bills and outdoor weather data to conduct a multiple-parameter regression. It can identify the hidden problem of main energy use systems, as well as other detailed information about physical properties of the building and cooling/heating system (i.e., cooling or heating load, the operation of cooling/heating system and the efficiency of cooling/heating system).

Until now, this diagnosis method has been adopted in 15 green office buildings, and good application results have been achieved. Some cases carried out further measures to improve energy conservation based on the model results, and the energy-saving rate after renovation even exceeded 20%.

- IEQ factor diagnosis model

Kano model was adopted to explore the subjective preference for IEQ factors, and to find out which IEQ factor has a relatively significant impact on the overall satisfaction. This method has been applied in a typical green office building, and two groups of occupants in private and open-plan offices were respectively investigated. Each IEQ factor was categorized as Basic, Proportional or Bonus factor, according to its relative magnitude of positive and negative impacts on the overall satisfaction. The results provided valuable suggestions on how to improve the IEQ performance in a cost-effective way.

References

1. Kissock JK, Haberl JS, Claridge DE (2002) Development of a toolkit for calculating linear, change-point linear and multiple-linear inverse building energy analysis models, ASHRAE research project 1050-RP. Texas A & M University, Final Report
2. Kim H (2012) Methodology for rating a building's overall performance based on the ASHRAE/CIBSE/USGBC performance measurement protocols for commercial buildings, Doctoral dissertation. Texas A&M University, College Station, TX
3. Kim K (2014) Development of an improved methodology for analyzing existing single-family residential energy use, Doctoral dissertation. Texas A&M University, College Station, TX
4. Kim KH, Haberl JS (2015) Development of methodology for calibrated simulation in single-family residential buildings using three-parameter change-point regression model. Energy Build 99:140–152
5. Kim H, Haberl J (2017) Field-test of the ASHRAE/CIBSE/USGBC performance measurement protocols: Part I intermediate level energy protocols. Sci Technol Built Environ
6. Kim H, Haberl J (2017) Field-test of the ASHRAE/CIBSE/USGBC performance measurement protocols: Part II advanced level energy protocols. Sci Technol Built Environ
7. Kissock JK, Eger C (2008) Measuring industrial energy savings. Appl Energy 85(5):347–361. https://doi.org/10.1016/j.apenergy.2007.06.020
8. Reichmuth H, Turner C (2010) A tool for efficient first views of commercial building energy performance. In: 2010 ACEEE summer study on energy efficiency in buildings, pp 325–338
9. Abels B, Sever F, Kissock K, Ayele D (2011) Understanding industrial energy use through lean energy analysis. SAE Int J Mater Manuf 4(1):495–504. https://doi.org/10.4271/2011-01-0326
10. GB/T (2015) Public building energy efficiency design standards. GB/T 50189-2015. Beijing: Ministry of Housing and Urban-Rural Development of the People's Republic of China (in Chinese)
11. Kano N (1984) Attractive quality and must be quality. J Jpn Soc for Qual Control 14(2):147–156
12. Kim J, de Dear R (2012) Nonlinear relationships between individual IEQ factors and overall workspace satisfaction. Build Environ 49:33–40
13. Kim J, Dear RD (2012) Impact of different building ventilation modes on occupant expectations of the main IEQ factors. Build Environ 57:184–193
14. Zagreus L, Huizenga C, Arens E, Lehrer D (2004) Listening to the occupants: a web-based indoor environmental quality survey. Indoor Air 14(Suppl. 8):65–74
15. Fanger PO (1972) Thermal comfort: analysis and applications in environmental engineering. McGraw-Hill, New York

Chapter 6
Further Exploration of Operating Performance of Individual Green Office Building Through Data Mining

6.1 Background

With the rapid development of big data and sensing technology, building energy consumption and IEQ data will be extensively accumulated during the operation stage. Take a typical office building as an example, it can easily produce millions of data related to energy consumption and IEQ in a year, which contain huge data mining values for the building performance diagnosis and improvement. However, conventional data analysis methods are generally inefficient to process such a large scale of data and cannot maximize the value of big data. More advanced and efficient data analysis methods must be adopted.

In this chapter, the author will use the new data mining technology to conduct in-depth building performance analyses based on the high-quality data. Two data mining cases will be studied. The first case is related to building load prediction and smart control of HVAC system, and the second case is the IEQ daily pattern clustering. The objective is to discover the hidden laws and knowledge inside the building performance data, and use the data mining results to better guide the building operation.

6.2 Building Load Prediction Towards Smart Control of HVAC System

6.2.1 Research Framework

Most of the HVAC control systems in office buildings are established on the feedback model, i.e. only taking actions after indoor environment problems have already occurred (such as increasing the cooling supply when indoor temperature

Y. Geng, *Data-Driven Research on Characteristics of Operating Performance of Green Office Buildings*, Springer Theses, https://doi.org/10.1007/978-981-95-2329-0_6

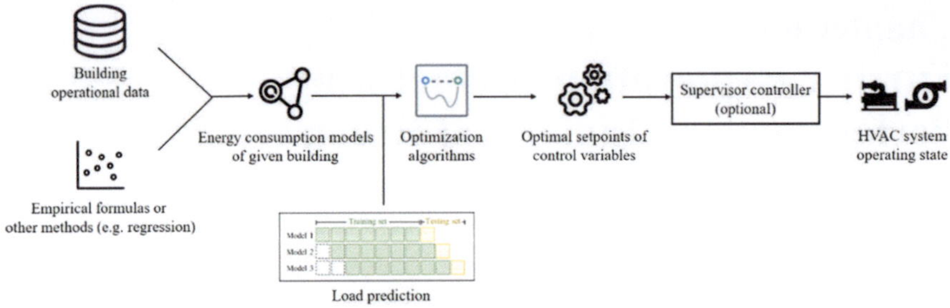

Fig. 6.1 The overall workflow of the predictive control method

has exceeded the upper bound of the comfort range in summer). This control logic may lead to the mismatch between supply and demand of air-conditioning, causing unnecessary energy waste. HVAC load forecasting can help buildings predict and cope with the changing load in advance. Therefore, the author proposed a predictive control method, which can solve the mismatch problem, bring energy saving effect and improve thermal comfort.

The overall workflow of the proposed method is shown in Fig. 6.1. First, a period of building operational data is collected, and then be used to build the energy consumption models based on empirical formulas or other methods. For different buildings, the forms of the models can be different. Then, the load prediction result is put into the model, and an optimization problem is defined. By solving this optimization problem, the optimal values (setpoints) of the variables are decided. Finally, the HVAC system will operate at the new optimal setpoints. Additionally, a supervisor controller can be set before the final step to decide if the optimal setpoints will be actually carry out in HVAC system operating. For example, if the new optimal setpoints is very close to current setpoints, the adjustment action will not be performed.

6.2.2 Case Information

The selected case is an office park (Fig. 6.2) located in Hangzhou, Zhejiang Province, China. This office park has 8 office buildings and includes 4 HVAC systems, and the author chose one HVAC system for the case study. The selected HVAC system provides the cooling for building #6, #7 and #8 in the park, with a total air conditioning area of more than 100,000 m^2. The building operational data collection platform and equipment management system of the office park are well-developed, which can obtain several types of historical data including equipment operation state, HVAC system parameters (such as chilled water flow rates and temperatures), occupant counts and outdoor weather parameters. The data sampling rate is 5 min. The building envelope is mainly glassing curtain wall, so solar radiation has a significant impact on the building cooling load.

Fig. 6.2 Photo of the selected case for the predictive control of HVAC system

The data is collected between June 1 and August 10. The prediction target at each prediction point is the cooling load in the next hour, and the history data is resampled every hour. Cooling load prediction was conducted out every hour, and the model was also retrained per hour. In every training process, the historical data within one week before the prediction point was used for training. The model randomly divides 70% of the training data into training set and the remaining 30% into the validation set. The test period was from August 2 to August 9.

The selected system includes four chillers, five chilling pump, five cooling pump and eight cooling towers. Key information of the HVAC equipment can be found in Table 6.1. The number of running pumps corresponds with the chillers. Since the cooling side equipment has fixed frequency, only the chilling side of this HVAC system is chosen as study case. The operational data is collected by the smart management platform installed in the office park, and the data sampling rate is set to one hour to avoid the chiller short cycling. The rated parameters of each equipment are shown in the tables below.

Table 6.1 The rated parameters of (a) chillers, (b) pumps, (c) cooling tower fans

(a)

Chiller type	Chiller count	Cooling capacity (kW)	Rated power (kW)	Rated COP
Large	3	4219.2	783	5.387
Small	1	3857	711	5.424

(b)

Pump type	Pump count	Rated flow rate (m^3/h)	Rated head (m)	Rated power (kW)	Frequency conversion
Chilling pump	5	850	41	132	TRUE
Cooling pump	5	1000	27	110	FALSE

(c)

Tower count	Rated flow rate (m^3/h)	Rated power (kW)	Frequency conversion	Rated frequency (Hz)
8	4800	18.5	False	50

6.2.3 Load Prediction Modelling and Results

The first step of the building load prediction modelling is to extract several features from the vast amount of historical data as the model inputs. According to the influence factors of building load, the features that require further particularized are determined. Building HVAC load mainly includes the heat loss through building envelope, the cooling loads required to handle fresh air, and internal heat gains (occupancy and electrical appliances inside the building), thus it is affected by outdoor weather conditions, occupancy, the type and operation of devices, and the property of building envelope.

Among these factors, the building envelope seldom changes after the building is constructed. In office buildings, the operation schedule of appliances is usually highly consistent, depending on the schedule of occupancy, so the number of people in the room is a good indicator of internal heat gains. In addition, the operation schedule of office buildings is relatively fixed and cyclical, so time-related features such as the hour of the day or the day of the week can well represent building operation schedule. The heat loss through building envelope mainly comes from the heat transfer from outdoor environment and ambient solar radiation, so the outdoor air temperature and solar radiation intensity are also included in the feature types. The ventilation load is affected by the enthalpy difference between indoor and outdoor air. However, the indoor environment is usually stable in air-conditioned period, so the humidity and enthalpy of outdoor air can be important features.

After determining the preliminary feature types, the final input features of the load prediction model can be decided according to the number of buildings in the HVAC system, the data quality and the load characteristics by analyzing the real data and identify the representativeness of each feature. Figure 6.3 shows the flow chart of the feature selection process.

Based on the analysis of real data, the input features of load prediction model were identified as Feature Set 1 (experiment group) in Table 6.2. Additionally, in order to validate the effectiveness of the proposed features, the author conducted a literature review to identify the features which are most used in existing studies for the building load prediction. Those features are named as Feature Set 2 (control group) in Table 6.2.

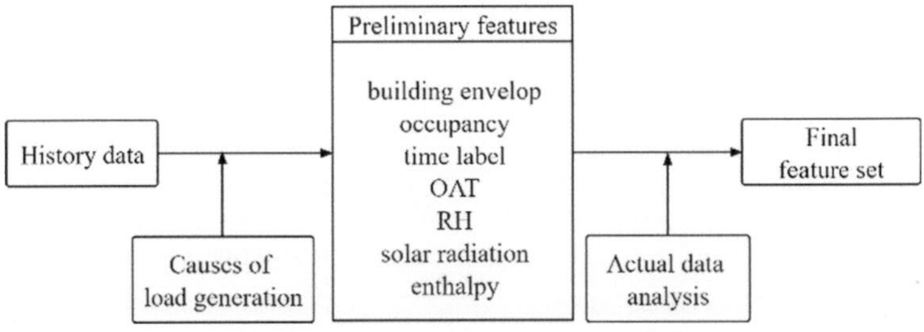

Fig. 6.3 Flow chart of the feature selection process for load prediction

Table 6.2 The features selected as model inputs

Features		Feature set 1	Feature set 2
Time label	Day of week	√	√
	Hour of day	√	√
	Whether is holiday	√	√
Black-bulb temperature	2nd floor on the west	√	×
	2nd floor on the east	√	×
	4th floor on the west	√	×
Occupancy	Number of people in building No.6	√	√
	Number of people in building No.7	√	√
	Number of people in building No.8	√	√
	Total number of people in all three buildings	√	√
Outdoor environment	Outdoor air temperature	√	√
	Outdoor air relative humidity	√	√
	Outdoor air enthalpy	√	×
Past load	Past 1 h cooling load	×	√

After the feature selection, an appropriate machine learning algorithm should be chosen for the modelling of load prediction. In this study, the author selected Light Gradient Boosting Machine (LightGBM). LightGBM is a gradient boosting framework that uses tree-based learning algorithms, which is usually used in time-series data prediction. Different from commonly used tree models, LightGBM grows trees leaf-wise in each decision tree. It will choose the leaf with max delta loss to grow instead of growing all leaves (as shown in Fig. 6.4), which makes the algorithm more accurate. It uses algorithms to find the best split point within each tree, which greatly reduces the time cost of model training while keeping the model accuracy basically unchanged.

The parameters of the model are critical to model performance. One model can have several parameters, and for different models, the best combinations of parameters vary a lot. Hence, the parameters should be tuned in each scenario based on

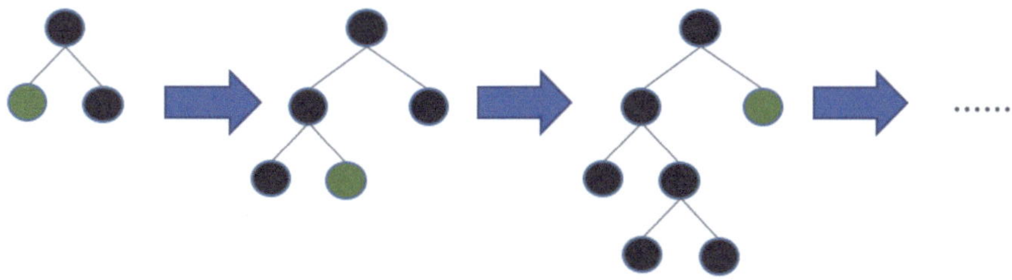

Fig. 6.4 Illustration of leaf-wise tree growth of LightGBM (Credit to LightGBM documentation)

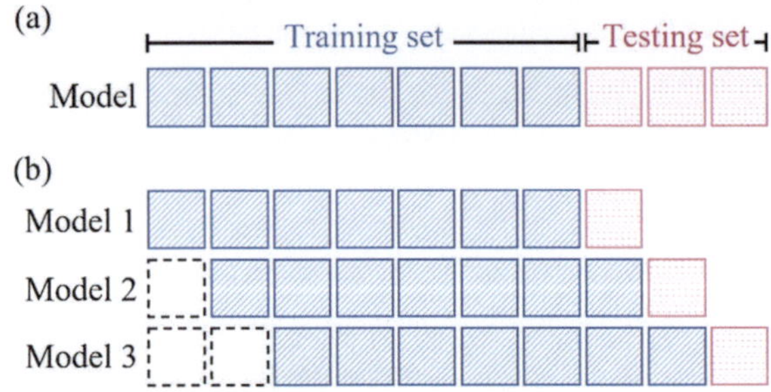

Fig. 6.5 Illustration of the training sets and test sets of **a** traditional prediction model and **b** rolling prediction model

historical data. In this study, an auto-tune tool named Optuna is used for searching the hyper-parameters in the model. It tries different hyper-parameter combinations within the specified parameter ranges to build and train models respectively, among which the model with the highest accuracy is selected as the best model, and the parameters of this best model are the optimal hyper-parameters.

The method of online learning is used for model training. Different from offline training which has fixed training set and testing set, online learning trains the model using the most recent data (as shown in Fig. 6.5), which means that the model needs to be regularly retrained. Although this approach leads to higher computational costs, it allows the model to always use the latest data and therefore being able to respond to exceptional situations in time such as abrupt weather changes.

In the process of parameter tuning and model training, coefficient of the variation of the root mean square error (CVRMSE) and mean absolute percentage error (MAPE) are used as the evaluation metrics of model accuracy. The two errors are calculated as follows:

$$CVRMSE = \frac{\sqrt{\frac{1}{n} \cdot \Sigma \left(y_i - \widehat{y_i}\right)^2}}{\overline{y}} \tag{6.1}$$

$$MAPE = \frac{1}{n} \cdot \Sigma \frac{\left|y_i - \widehat{y_i}\right|}{\widehat{y_i}} \tag{6.2}$$

n the number of data
yi the ith term of the predicted data
$\widehat{y_i}$ the ith term of the ground truth data
$\overline{y}$ the average of the ground truth data.

Figure 6.6 shows the load prediction results during the test period. Both load prediction models using Feature Set 1 and 2 show a good prediction accuracy in general. However, the prediction result by the model using Feature Set 2 generally failed to respond in time and leads to large prediction errors when the cooling load changed rapidly. The main reason is that Feature Set 2 used the historical cooling load in the previous hour as one input. This model input played a dominant role in

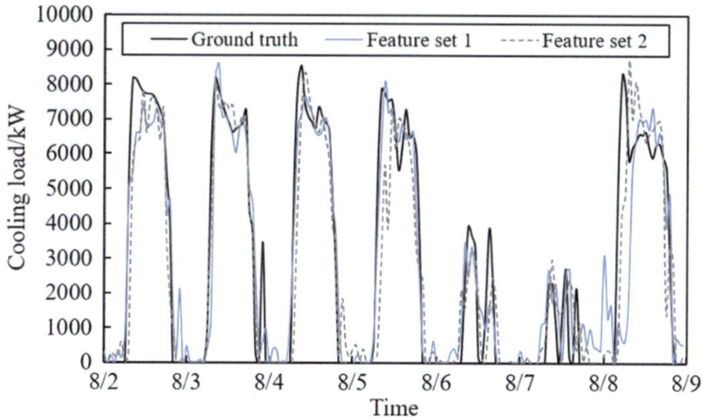

Fig. 6.6 Cooling load prediction results during the test period

the load prediction, which made the predicted load always tend to be close to the load at the last hour.

Since the load prediction model is used to inform the optimal operation of the HVAC system, in addition to the errors during the whole test period, the errors during the system operating period should also be concerned. The evaluation metrics during the whole test periods (CVRMSE-all, MAPE-all) and system operating periods (CVRMSE-on, MAPE-on) on the models are shown in Table 6.3. It can be found that the CVRMSE-all of model with the Feature Set 1 is 6% lower than that of model with the Feature Set 2, while the CVRMSE-on is 21% lower. The MAPE-all and MAPE-on also decreased by 9% and 30%. This indicates that the model using the proposed feature set outperforms the model using common features, especially for the load prediction in operation periods, demonstrating the rationality and effectiveness of the proposed feature selection method.

Table 6.3 Evaluation metrics on load prediction models with the two feature sets

	Model with feature set 1	Model with feature set 2
CVRMSE-all	0.392	0.415
MAPE-all	0.280	0.309
CVRMSE-on	0.220	0.280
MAPE-on	0.137	0.197

6.2.4 Control Strategy Modelling and Results

To achieve the required indoor environment, the chilled water supply temperature, the frequency of the pumps and fans need to be adjusted. Therefore, the relationships between these control variables and the energy consumption of corresponding equipment need to be found. With the building operational data obtained, the multiple linear regression (MLR) method is used to fit the energy consumption and the control variables.

For chillers, the empirical formula provided by ASHRAE handbook [1] is used, since the cooling load (Q_{ch}) can be known by load prediction model. The formula is shown below, where P_{ch} is the power rate of the chiller, t_{cwr} is the cooling water return temperature, t_{chws} is the chilled water supply temperature, and a_0 to a_5 are regression coefficients fitted based on building history operational data.

$$P_{ch} = a_0 + a_1(t_{cwr} - t_{chws}) + a_2(t_{cwr} - t_{chws})^2 + a_3Q_{ch} + a_4Q_{ch}^2 + a_5(t_{cwr} - t_{chws})Q_{ch} \tag{6.3}$$

Figure 6.7 shows the chiller power fitting result under four operation conditions. The chiller power calculated by the model (Eq. 6.3) is compared with the measured power to evaluate the accuracy of the model. The average R square of the four conditions is 0.84, showing that the model is reliable for estimating the power of the chillers. The operation conditions with less data have relatively smaller R square value, indicating that larger data amount is beneficial for model accuracy. The values of the regression coefficients are also listed in Table 6.4.

For pumps and fans, the calculation methods are different between fixed and variable frequency devices. For fixed frequency pumps and fans, the pump power can be estimated as the rated power. For variable frequency pumps and fans, the energy consumption is relative to the flow rate and the frequency, and can vary in different pipeline resistance conditions. Usually, the energy consumption model for variable frequency pumps or fans can be simplified as quadratic or cubic equation of the flow rate [2, 3], but during the operation of the HVAC system, the pipeline resistance is changing with the running state of air-condition terminals, and the energy consumption is hard to be described by a fixed equation. However, the frequency needed to achieve the required flow rate can indirectly show the pipeline resistance. So, in this study, the energy consumption of variable speed pump or fan can be simplified as a cubic homogeneous equation of the flow rate and frequency as below, where P_p is the power rate of the pump or fan, f is the frequency, L is the flow rate, and b_0 to b_9 are regression coefficients.

$$P_p = b_0 + \mathrm{b}_1f + b_2L + b_3fL + b_4f^2 + b_5L^2 + b_6f^2L + b_7fL^2 + b_8f^3 + b_9L^3 \tag{6.4}$$

Through validating the model by actual building operational data, it was found that the regression coefficients of the cubic terms (b_8 and b_9) are too small to affect the

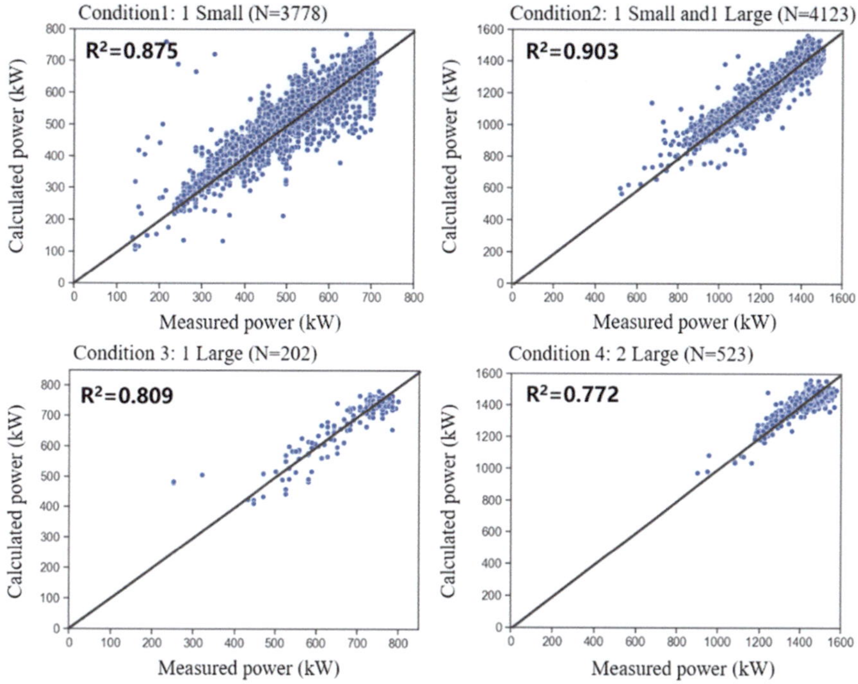

Fig. 6.7 The fitting result and R square of the chiller power in four operation conditions

Table 6.4 The regression coefficients of the chiller energy consumption models

Chiller operation condition	a_0	a_1	a_2	a_3	a_4	a_5
1 Small	251.86	− 23.33	0.90	0.05	1.13e−05	0.0016
1 Small and 1 Large	1624.24	− 129.08	3.28	0.09	5.78e−06	− 0.0003
1 Large	1257.74	− 79.47	0.50	0.05	− 3.62e−05	0.0152
2 Large	1679.24	− 21.34	− 0.79	− 0.21	6.55e−06	0.0107

result, so the cubic terms were omitted in order to simplify the model. Furthermore, through the on-site test of the HVAC system, it is discovered that the flow rate and power of the pump can be described as a linear relationship under the same frequency, as shown in Fig. 6.8. The operational data also shows the same trend. Therefore, the power model of the pump can be further simplified by omitting the secondary form of the flow rate (b_5 and b_7). The final model used in this study is shown as below:

$$P_p = b_0 + b_1 f + b_2 L + b_3 fL + b_4 f^2 + b_6 f^2 L \tag{6.5}$$

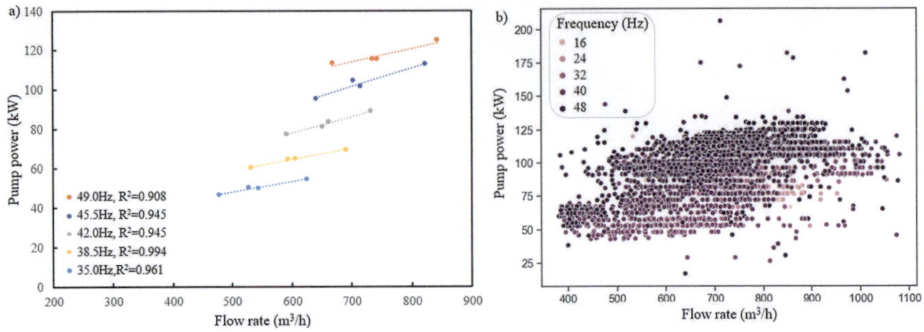

Fig. 6.8 The flow rate and power of the pump in **a** on-site test result, **b** history operational data

Figure 6.9 shows the fitting result of the pump power. According to the result, the pump power model has enough accuracy. The overall R square is 0.753, and the absolute error is mainly within 25 kW. The coefficients are also listed in Table 6.5.

The optimal setpoints of the control variables are found by solving the optimization problem. The objective function is minimization of the total power of the HVAC system, which is the total power of each equipment. Based on the models mentioned above, the total power can be described as a function of the control variables below.

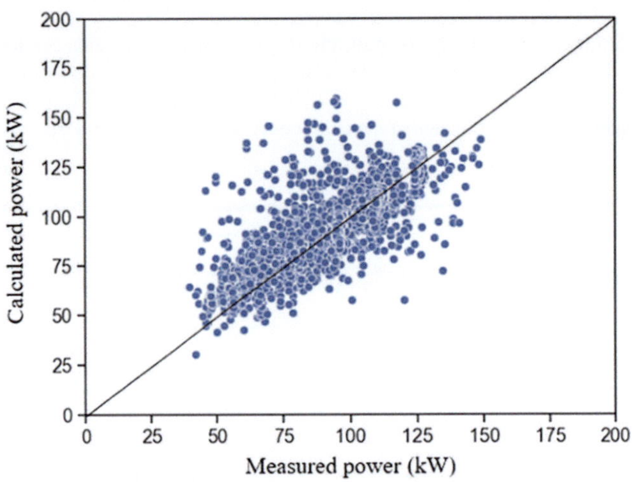

Fig. 6.9 The fitting result of the pump power

Table 6.5 The regression coefficients of the pump energy consumption model

Coefficient	b_0	b_1	b_2	b_3	b_4	b_6
Value	139.795	– 0.014	– 8.421	0.132	0.005	– 5.58e–05

Table 6.6 The lower and upper limits of each control variable

Control variable	Lower limit	Upper limit
t_{chws} (°C)	9	18
t_{chwr} (°C)	–	24
Δt_{chw} (°C)	–	8
f_{pump} (Hz)	35	50

$$obj = \min\left(P_{ch} + P_{pump} + P_{fan}\right) = \min \mathrm{F}\left(\mathrm{t_{chws}},\ \mathrm{t_{cwr}},\ \mathrm{f_{pump}},\ \mathrm{f_{fan}},\ \mathrm{L_{pump}},\ \mathrm{L_{fan}}\right) \tag{6.6}$$

There is also a constraint between the frequency and flow rate of a single pump or fan. Under the same pipeline resistance, the flow rate and frequency should be proportional. In this study, the most recent recorded history data is used to determine the pipeline resistance.

$$\frac{L_{pump}}{f_{pump}} = \frac{L_{pump_{last}}}{f_{pump_{last}}} \tag{6.7}$$

Furthermore, Table 6.6 shows the lower and upper limits of the control variables, which are determined based on the domain knowledge. With the objective function and the constraints defined, the optimization problem can be solved within a certain search domain, which is set according to the actual situation.

Finally, the author selected a certain optimization algorithm to find the optimal setpoints of chilled water supply temperature and chilling water pump frequency. Based on the optimization result of control variables, the corresponding energy cost can be calculated as well.

The algorithms for solving optimization problems mainly include traditional optimization algorithms and heuristic optimization algorithms. Traditional optimization algorithms include gradient descent method, Newton's method, conjugate gradient method, etc. They have fixed structures and parameters and are generally deterministic algorithms. Heuristic optimization algorithms include genetic algorithms (GA), particle swarm algorithms (PSO), etc. They can adaptively adjust the search direction, and have better global search capability. In this study, sequential least squares programming (SLSQP) and GA are used as representatives of two types of optimization algorithms.

SLSQP uses Taylor expansion to transform the constraint problem to a quadratic programming problem. SLSQP has the advantages of good convergence and high computational efficiency. GA is a computational model that simulates biological evolution and genetics to search for optimal solutions by simulating natural evolutionary processes. GA tends to obtain better results when solving more complex optimization problems and are more likely to find the global optimal solution. The optimization results obtained by these two algorithms were further compared.

The two optimal control variables (chilled water supply temperature and pump frequency) on June 29 were shown as an example in Fig. 6.10. The optimal t_{chws}

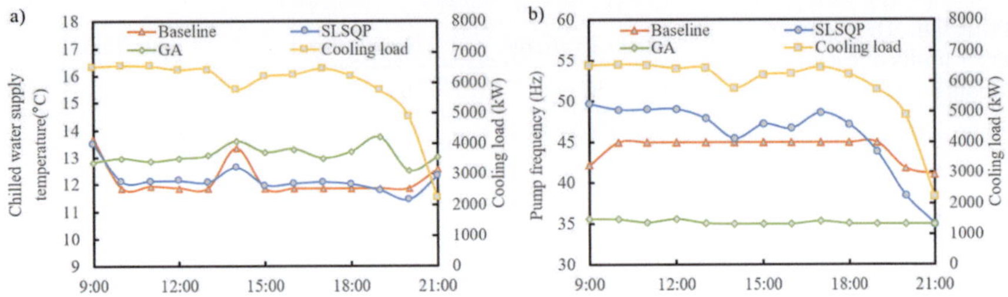

Fig. 6.10 Chilled water supply temperature and pump frequency of the baseline and two optimization results

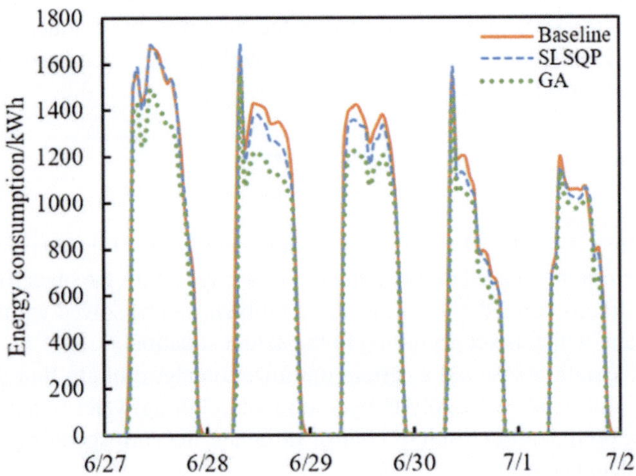

Fig. 6.11 The energy consumption of the case system before and after optimization

found by SLSQP is similar to that in baseline condition, while GA gives a higher value. However, the pump frequency given by GA is notably lower than the baseline and SLSQP. It indicates that GA optimization suggests a higher chilled water return temperature, which may not be feasible in some of the actual control scenarios, since the cooling capacity of the terminals may be reduced under higher chilled water temperature. Therefore, in practical application, it is suggested to determine the search domain based on the performance parameters of the terminals. Contrarily, SLSQP optimization suggests the operation mode of higher water flow rate and a bit higher water temperature, which can increase the COP of the chiller and achieves energy saving. Due to the manual-set initial point, the optimal setpoints given by SLSQP is more conservative, but still have certain energy saving effect.

Figure 6.11 shows the energy use before and after optimization. The result indicates that both optimization algorithms can achieve a certain energy saving effect. The optimization model provides 4.52% energy saving by SLSQP, and 11.17% by GA, indicates that GA can outperform SLSQP in this case.

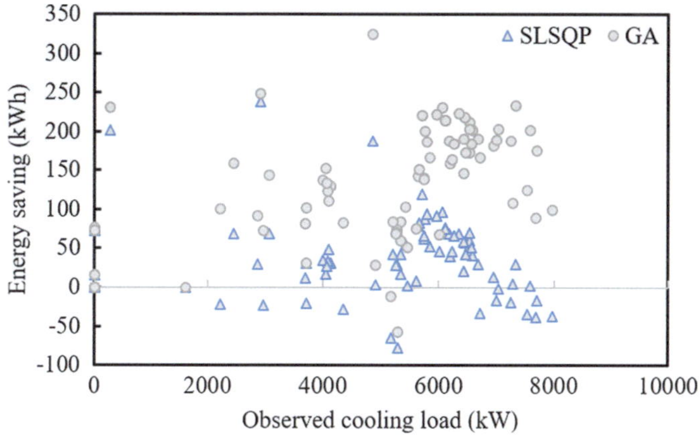

Fig. 6.12 Scatter plot of the cooling load and the energy saving effect

Figure 6.12 shows that the energy saving effect varies with the cooling load. The highest energy saving happens when the cooling load is between 5500 and 6500 kW, where the cooling load has not reached the maximum cooling capacity, and two chillers are running. However, when the cooling load is near maximum (beyond 7500 kW), the HVAC system needs full load operation, and energy saving effect becomes lower.

It is worth mentioning that under maximum cooling load, SLSQP nearly has no energy saving effect, but GA can still provide certain energy saving. This is because SLSQP is a quasi-Newton method, and relies on a manually set initial point as the start of optimal searching. Different initial points can greatly affect the optimization result. To ensure the stability of the optimization process, the initial point is set near the baseline parameters, which causes the optimal setpoints closer to the baseline, reducing the energy saving effect. However, GA generates the initial 'parents' by random sampling, which is more likely to find the global optimal solution. This can also explain the performance gap between the two algorithms.

6.3 IEQ Daily Pattern Clustering for Building Operation Diagnosis

6.3.1 Research Framework

IEQ data has a high spatial–temporal resolution and dimension, which requires exploring the rich values. Identifying typical IEQ patterns is one of the most essential pieces of knowledge discovered in big data [4]. In some studies, pattern recognition has been considered an effective way to understand the characteristics of daily IEQ

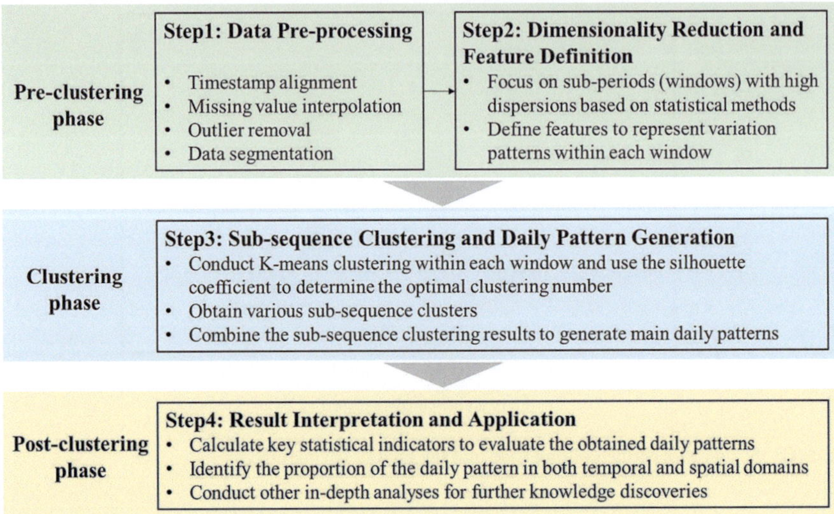

Fig. 6.13 The IEQ clustering framework proposed in this study

variations [5]. Furthermore, researchers can perform anomaly detection and operation management after obtaining typical IEQ patterns [6]. For example, building operators can determine the possible unreasonable HVAC operation mode based on the obtained daily patterns of indoor air temperature, and then make targeted adjustments to avoid energy waste. Therefore, the potential application values of IEQ pattern identification are remarkable.

This study used the clustering method to extract inherent IEQ patterns for further building performance evaluation and diagnosis. Figure 6.13 shows the proposed IEQ clustering framework. The overall workflow consisted of three phases (pre-clustering, clustering, and post-clustering phase). Data pre-processing, dimensionality reduction, and feature definitions were conducted in sequence in the pre-clustering phase. Sub-sequence clustering was performed and the main daily patterns were generated in the clustering phase. Finally, the obtained clustering results were interpreted and other applications were explored (i.e., IEQ evaluation or diagnosis) based on secondary analyses and domain knowledge.

6.3.2 IEQ Clustering Difficulty Analysis and Solutions

According to previous time-series clustering analyses, data dimensionality reduction is the key step that determines the clustering performance. Considering a set of time-series monitoring data with a 1 min interval, the dataset theoretically has 1440 dimensions per day. If the high-dimensional raw data are clustered without dimensionality reduction, many clustering algorithms will become ineffective because of the invalidity of the distance function. This problem is called the "Curse of Dimensionality" [7]. Thus, dimensionality reduction is necessary before time-series clustering.

However, current dimensionality reduction and clustering methods are mostly specific to energy data and not appropriate for the IEQ. There are two main reasons: (1) Generally, IEQ data often have a higher dimension than energy data. The higher dimension of IEQ data gives the more challenging task of minimizing information loss when conducting data dimensionality reduction. (2) IEQ is more sensitive to human activity and other factors. Thus, its variation may be more frequent and complex than the variation in energy data. In particular, IEQ data may have more short-term fluctuations, which are important but difficult to extract using traditional clustering methods.

Faced with the above challenges, this study aims to develop a new time-series clustering method for IEQ daily pattern extraction and apply the method to real building cases for validation and further IEQ evaluation and diagnosis. The new method should achieve a good balance between dimensionality reduction and fine characterization (especially the extraction of local variation characteristics in specific short periods).

After a series of considerations and attempts, the concept of "sub-sequence" is introduced into the step of dimensionality reduction and clustering, because it is unnecessary to consider the entire day in IEQ clustering cases. For example, in an office building, the indoor air temperature generally varies significantly around the working start time or leaving time, whereas it remains relatively stable during the working time. Hence, a long period without significant environmental changes should be excluded from further feature definitions and clustering.

The sub-sequences indicate some significant "windows" (generally lasting for a few hours), during which the IEQ data are most likely to experience significant variation or have differential characteristics. In this study, the variation of the environmental parameter per 0.5 or 1 h was calculated, as well as the standard deviation of variation among all daily curves. These indicators help select the noteworthy sub-sequences from the daily profile, which will be further described in the following case studies. The selected sub-sequences can help exclude other inessential periods in the daily profile, which reduces the data dimension and allows focus on more short-term variation characteristics within the most critical sub-periods.

Several features were then defined to represent the local variation pattern within each window. The segment average method for feature definition was used to avoid destroying the original time-series data structure and nature. However, the limitation of information loss by averaging can be overcome because the sub-sequence has already compressed the time range, which means that a limited number of features can be used to capture the short-term variation with a higher temporal resolution (less than one hour or half hour). For each sub-sequence, average segment values cannot be directly used as the clustering features due to the shape similarity. Therefore, the differences between the average values of each pair of adjacent segments were calculated, and used to reflect the shape of the IEQ variation. It is worth noting that changing the distance measure during the clustering (for example, using cosine distance to replace Euclidean distance) can have a similar effect. Nevertheless, the transformation of feature definition is more intuitive and easily implemented.

After the dimensionality reduction and feature definition, sub-sequence clustering was performed within each window. The selected clustering algorithm was k-means, owing to its easy implementation and high efficiency. The k-means clustering procedure is as follows: (1) randomly select k points as the initial centers; (2) assign each point to the closed cluster based on the minimum Euclidean distance; (3) update each cluster's center with the average of all sample points in that cluster; (4) repeat Step 2 and 3 until each cluster's center does not change (which indicates the algorithm converges); and (5) output the final cluster centers and k clusters.

Since the initial centroid selection is random, it can have a great influence on the final clustering result. To identify the optimal initial centroids and the corresponding clustering result, 50 iterative computations (learned from a previous study conducted by Liu [8]) were performed for every clustering number ranging from 2 to 10, respectively. The silhouette score [9] was calculated for each clustering computation. Finally, the clustering number and result with the highest silhouette score were saved. Based on the clusters obtained in each window, permutation and combination can easily generate the entire daily pattern.

6.3.3 *Case Study: Temperature Clustering in an Office Building*

The first case building to examine the performance of the proposed clustering framework is an office building in Shanghai, China. It is a two-story building converted from a factory plant. The total floor area was 1556 m^2 with 216 working stations. Furthermore, a variable refrigerant volume (VRV) system was used for heating and cooling.

During the measurement period in summer, 62 temperature sensors were installed in the main open office area (as shown in Fig. 6.14) to continuously monitor the indoor thermal performance with an interval of 1 min. Figure 6.15 shows the outdoor air temperature and relative humidity during the measurement period. Consequently, 868 raw daily curves of indoor air temperature were obtained. After data pre-processing, 734 daily curves were chosen for clustering analysis. They consisted of measurements from 62 temperature sensors and the proportion of each sensor was almost equal.

Figure 6.16 shows the distribution and standard deviation of temperature variation per 30 min among 734 daily curves. It can be seen that 7:30–9:30 and 17:00–21:30 had relatively higher dispersions (standard deviation > 0.3 °C) than others. Although some sub-periods within 17:00–21:30 did not have significant differences, it was unnecessary to break it up further considering its short period length. Therefore, we only focused on 7:30–9:30 (window 1) and 17:00–21:30 (window 2) to conduct sub-sequence clustering.

a)
b)

Fig. 6.14 **a** Photograph of an open office area in the office building; **b** floor plan and the sensor locations (each blue dot represents a sensor)

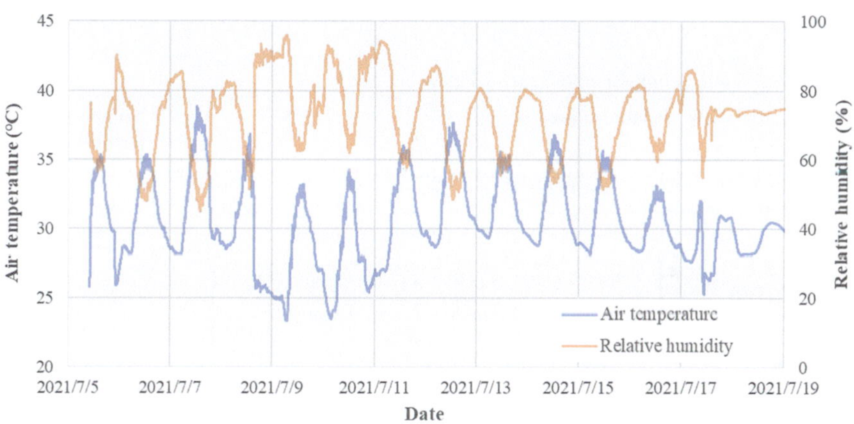

Fig. 6.15 Outdoor air temperature and relative humidity during the measured period

Table 6.7 presents the defined features in window 1 and 2. First, the average value of the air temperature every half hour was calculated. Second, the temperatures at adjacent time stamps were subtracted to exclude the trend similarity. As a result, each window's short-term (half-hour) variation characteristics in each window can be described with a few features (less than 10).

After the feature definition, sub-sequence clustering using the k-means algorithm was performed for each window. The optimal cluster numbers in windows 1 and 2 are 4 and 2, respectively. Table 6.8 presents the sub-sequence clustering results.

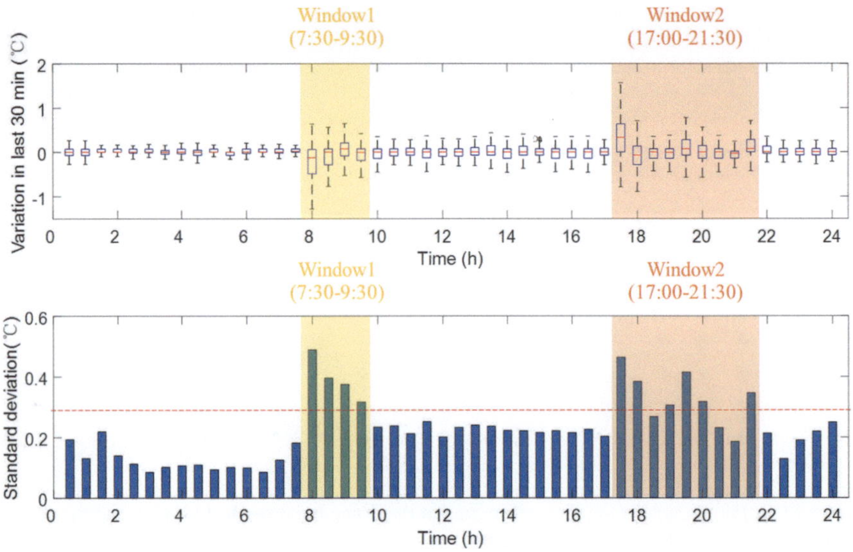

Fig. 6.16 The selection of sub-sequences in Case 1

Table 6.7 Feature definition within each window in Case 1

Clustering window	Feature subset	Number of features
Window 1 (7:30–9:30)	$\left\{\begin{array}{l}\overline{T_{8:00}} - \overline{T_{7:30}}, \overline{T_{8:30}} - \overline{T_{8:00}} \\ \overline{T_{9:00}} - \overline{T_{8:30}}, \overline{T_{9:30}} - \overline{T_{9:00}}\end{array}\right\}$	4
Window 2 (17:00–21:30)	$\left\{\begin{array}{l}\overline{T_{17:30}} - \overline{T_{17:00}}\ \overline{T_{18:00}} - \overline{T_{17:30}}\ \overline{T_{18:30}} - \overline{T_{18:00}} \\ \overline{T_{19:00}} - \overline{T_{18:30}}, \overline{T_{19:30}} - \overline{T_{19:00}}, \overline{T_{20:00}} - \overline{T_{19:30}} \\ \overline{T_{20:30}} - \overline{T_{20:00}}, \overline{T_{21:00}} - \overline{T_{20:30}}, \overline{T_{21:30}} - \overline{T_{21:00}}\end{array}\right\}$	9

For window 1 (7:30–9:30), four sub-clusters represented different indoor air temperature variation characteristics in the early morning. *Cluster 1–1* remains relatively stable during the sub-period, while the other sub-clusters exhibit a sharp drop and remain steady. The main difference among *Clusters 1–2*, *1–3* and *1–4* is the temperature dropping time. *Cluster 1–2* starts to decrease between 7:30 and 8:00; *Cluster 1–3* starts to decrease after 9:00; and *Cluster 1–4* starts to decrease at approximately 8:30.

For window 2 (17:00–21:30), two sub-clusters were obtained, which described different temperature characteristics at night. *Cluster 2–1* keeps stable; and *Cluster 2–2* experiences a periodic fluctuation, which means the extra work may occur at night.

After obtaining the sub-sequence clustering results, they are combined to generate all the daily patterns. In theory, 8 daily patterns (4 × 2) can be obtained, while

Table 6.8 Sub-sequence clustering results in Case 1

Clustering window	Sub-sequence clustering result
Window1 (7:30–9:30)	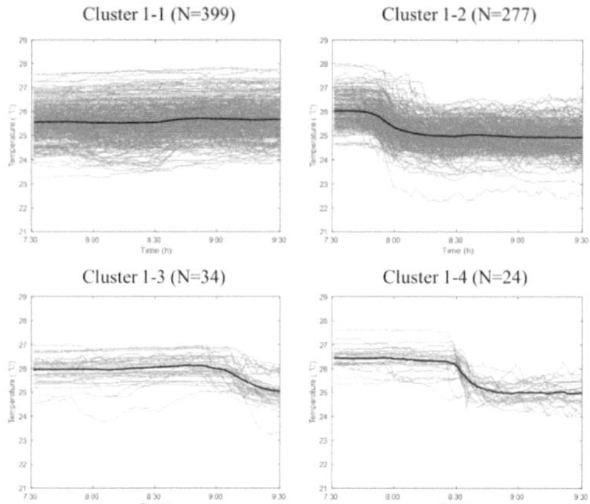
Window2 (17:00–21:30)	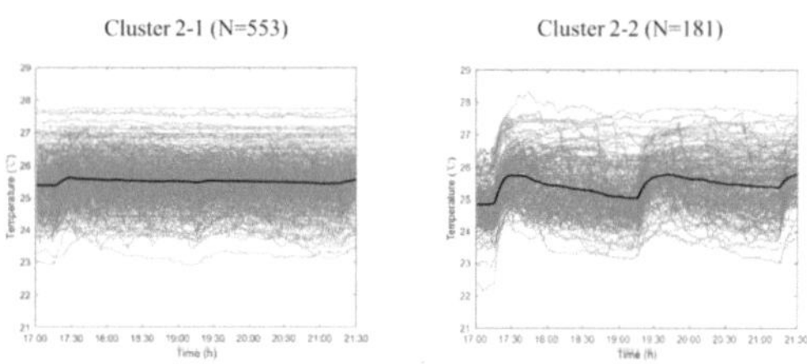

only three daily patterns account for more than 5% of the total dataset. Figure 6.17 illustrates the proportion of each daily cluster and further displays the profiles of the three main clusters. *Cluster 1* remained steady throughout the day, with the greatest number of curves. However, the mean indoor air temperature in *Cluster 1* varied significantly (from 23 to 28 °C). This implies that there may be different operational modes in *Cluster 1*, which will be further discussed. *Clusters 2* and *3* had a similar diurnal pattern, first experiencing a cliff descent in the early morning and then remaining steady until 17:30, while the temperature variation pattern became different after 17:30. The nocturnal variation in *Cluster 3* was more frequent and intense than that in *Cluster 2*.

A series of post-clustering analyses were conducted to understand the clustering result further and apply it to practice. Figure 6.18 shows the distribution of the occurrence days in a week for each cluster. It can be seen that *Cluster 1* frequently occurs on both weekdays and weekends, and the single-day proportion on weekends is relatively higher; while *Cluster 2* and *3* occur almost always on weekdays. The universality of *Cluster 1* reflects that there may exist two situations: (1) the HVAC system is off all day and the indoor air temperature is relatively high, which is likely

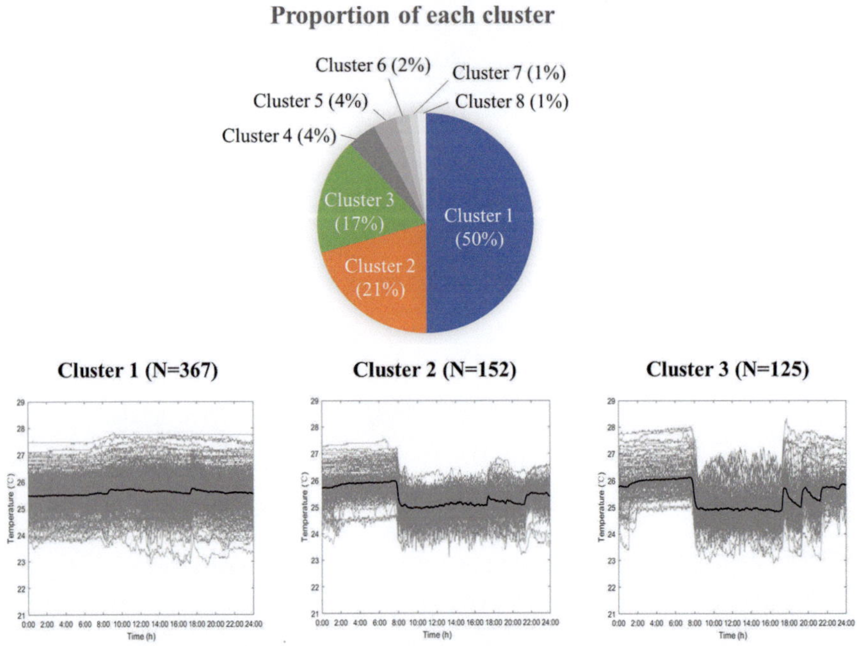

Fig. 6.17 The proportion of each daily cluster and typical patterns of indoor air temperature in Case 1

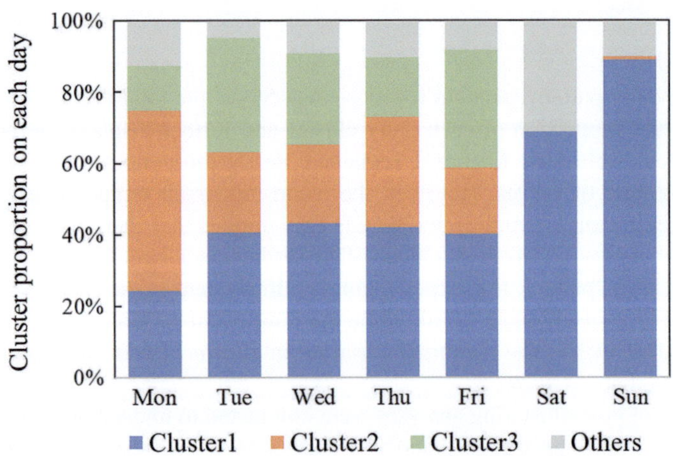

Fig. 6.18 The distribution of the occurrence days in a week for each cluster in Case 1

to occur on weekends; and (2) the HVAC system is on all day and the temperature remains relatively low throughout the day, which is likely to occur on weekdays.

After identifying the temporal distribution of daily clusters, their spatial distribution is further explored, as shown in Fig. 6.19. Considering that *Cluster 1* on weekdays and weekends may correspond to different operational modes, the author separated *Cluster 1* to determine where it occurred on weekdays and weekends.

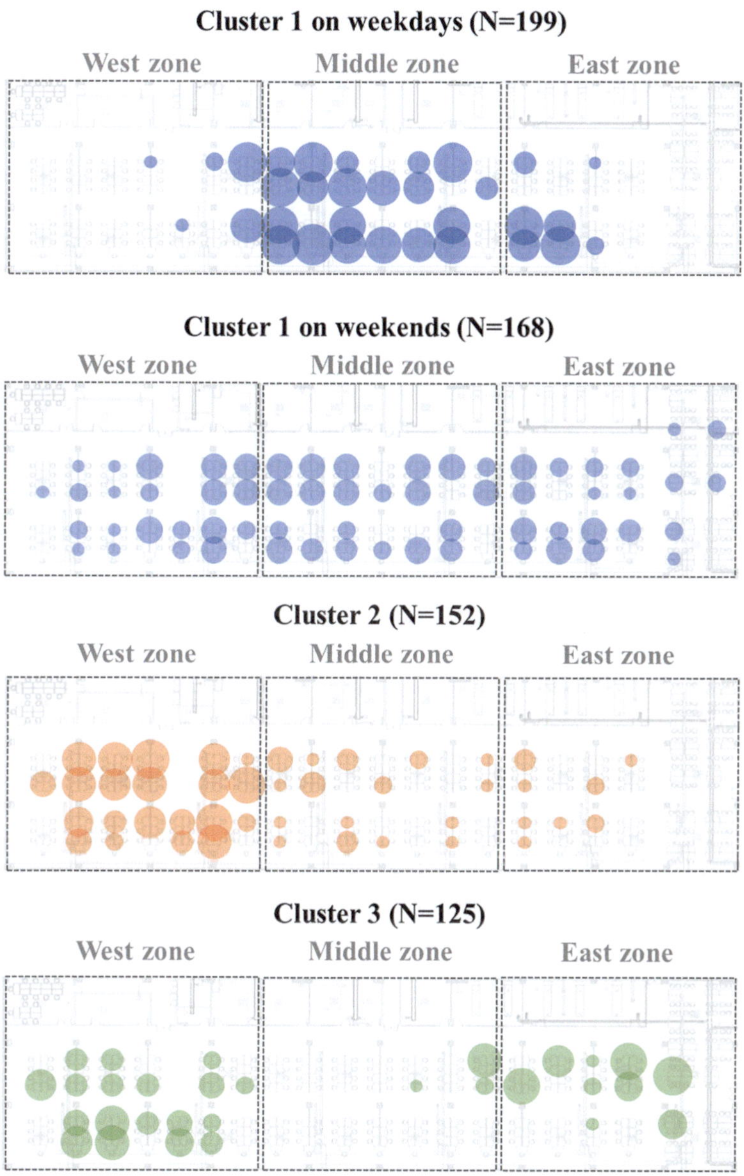

Fig. 6.19 The spatial distribution of the main daily clusters in Case 1 (The size of the bubble represents the occurrence frequency on each place)

Table 6.9 Input variables in the decision tree model

Variable	Type	Description
Day	Categorical	*Monday, Tuesday, Wednesday, Thursday, Friday, Saturday*, and *Sunday*
Zone	Categorical	*West, Middle*, and *East*
Outdoor_Temp	Numerical	Daily average outdoor air temperature (°C)
Outdoor_RH	Numerical	Daily average outdoor air humidity (%)

According to Fig. 6.19, *Cluster 1* on weekdays mainly occurs in the middle of the open office, which means that some HVAC systems in the central area continue to operate most of the time on weekdays. *Cluster 1* on weekends can be found throughout the space, meaning that almost all HVAC systems are turned off on weekends. *Cluster 2* and *3* mostly occur in the west or east of the office, which means that people in these areas often have extra work at night.

To further understand the rule of potential factors on different daily patterns, a classification model was developed using CART algorithm. The potential factors (model inputs) considered in this study are summarized in Table 6.9. To reduce the complexity of the decision tree and avoid overfitting, the minimum number of samples for a node split (*minsplit*) was set to 20. Furthermore, the post-pruning process was conducted by setting a cost complexity (*cp*) parameter, which was determined through the one-standard error in this study.

The classification tree is visualized in Fig. 6.20 and there are five terminal nodes, representing five decision rules (shown in Table 6.10). It should be noted that although we selected the outdoor temperature and relative humidity as the model inputs, they are not used in the classification tree. The occurrence frequency of different clusters (*Clusters 1, 2, 3* and *others*) is also given in Table 6.10. The highest frequency is marked in bold. Based on the obtained decision rules, it is possible to predict an expected pattern with a high probability under a given boundary condition. For example, when the "Day" is weekend, the indoor temperature probably follows the pattern of *Cluster 1* because the occurrence frequency of *Cluster 1* reaches 79% under this boundary condition. However, when predicting the western profiles on Tuesday, Wednesday and Friday ("Day = Tue, Wed or Fri" and "Zone = West"), and the eastern profiles on weekdays ("Day = Mon, Tue, Wed, Thu or Fri" and "Zone = East"), the probability can be low (47% and 38%). This indicates that the occupancy schedule and HVAC operational mode in these areas and days can be stochastic and uncertain.

6.4 Summary

This chapter presents two applications of data mining technology for the deep building performance analyses.

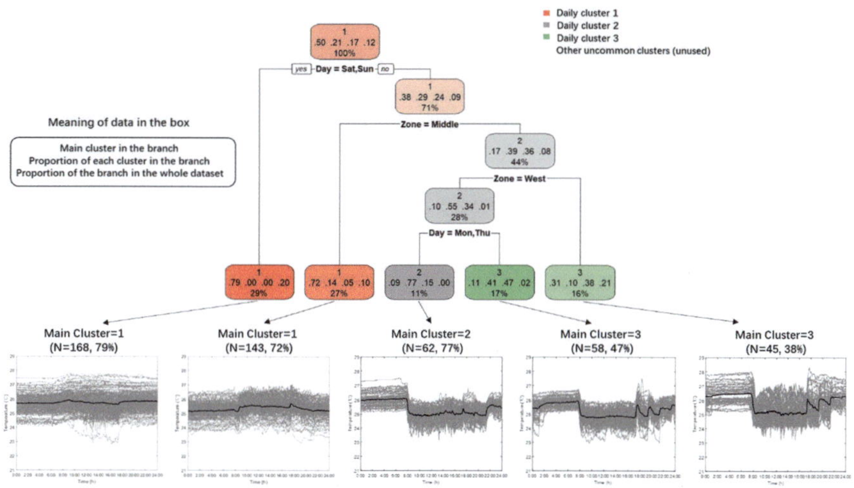

Fig. 6.20 Classification tree developed for the further knowledge discovery of the daily patterns in Case 1

Table 6.10 Decision rules of daily temperature patterns generated by the classification tree

Decision rule	Occurrence frequency of each cluster	Main cluster
If Day = Sat or Sun	**79%**, 0%, 0%, 20%	1
If (Day = Mon, Tue, Wed, Thu or Fri) AND (Zone = Middle)	**72%**, 14%, 5%, 10%	1
If (Day = Mon or Thu) AND (Zone = West)	9%, **77%**, 15%, 0%	2
If (Day = Tue, Wed or Fri) AND (Zone = West)	11%, 41%, **47%**, 2%	3
If (Day = Mon, Tue, Wed, Thu or Fri) AND (Zone = East)	31%, 10%, **38%**, 21%	3

Case 1 is the building load prediction and control strategy optimization of HVAC system. Firstly, a new feature selection method for the short-term HVAC load prediction was proposed. The load prediction model with the new feature set can reduce the prediction CVRMSE by 21% and MAPE by 30% during the system operating period, compared with the model with the original feature set. Secondly, a method for the dynamic optimization of HVAC system control variables at a given load was proposed. Two representative optimization algorithms (SLSQP and GA) were selected. Compared to the manual control strategy, the smart control strategy with SLSQP and GA saved 5% and 12% energy costs, respectively. The above results can help develop smart HVAC control system, which can improve the level of operation management and energy efficiency significantly.

Case 2 is the IEQ daily pattern clustering for building operation diagnosis. A new sub-sequence clustering framework for IEQ daily pattern extraction was proposed. The main advantage of the new clustering method compared with other time-series clustering methods is its ability to balance dimensionality reduction and fine characterization. The sub-sequence clustering method can exclude inessential periods without IEQ variations and only focuses on finer short-term variation characteristics within the most important periods. Therefore, it can significantly reduce the data dimension and simultaneously retain the original information of high-dimension data. Based on the clustering results and a series of post-clustering analyses, the author not only recognizes the IEQ variation characteristics in a day, but also finds out unreasonable building operation patterns and their main occurrence times or places.

References

1. ASHRAE (2019) ASHRAE handbook of heating, Ventilation, and air-conditioning applications, Atlanta, Georgia
2. Yan C, Wang S, Xiao F et al (2015) A multi-level energy performance diagnosis method for energy information poor buildings. Energy 83:189–203
3. Zeng Y (2014) Research on dynamic control of chilled water based on chilled model and load forecasting in air-conditioning system. Master thesis, Xi'an University of Architecture and Technology (in Chinese)
4. Miller C, Nagy Z, Schlueter A (2015) Automated daily pattern filtering of measured building performance data. Autom Constr 49:1–17. https://doi.org/10.1016/j.autcon.2014.09.004
5. Pereira PF, Ramos NMM, Ferreira A (2020) Room-scale analysis of spatial and human factors affecting indoor environmental quality in Porto residential flats. Build Environ 186:107376. https://doi.org/10.1016/j.buildenv.2020.107376
6. Yu Z, Haghighat F, Fung BCM (2016) Advances and challenges in building engineering and data mining applications for energy-efficient communities. Sustain Cities Soc 25:33–38. https://doi.org/10.1016/j.scs.2015.12.001
7. Verleysen M, François D (2005) The curse of dimensionality in data mining and time series prediction. In: Cabestany J, Prieto A, Sandoval F (eds) Computational intelligence and bioinspired systems. Springer, Berlin, Heidelberg, pp 758–770. Print ISSN: 978-3-540-26208-4, https://doi.org/10.1007/11494669_93
8. Liu X, Ding Y, Tang H, Xiao F (2021) A data mining-based framework for the identification of daily electricity usage patterns and anomaly detection in building electricity consumption data. Energy Build 231:110601. https://doi.org/10.1016/j.enbuild.2020.110601
9. Rousseeuw PJ (1987) Silhouettes: a graphical aid to the interpretation and validation of cluster analysis. J Comput Appl Math 20:53–65. https://doi.org/10.1016/0377-0427(87)90125-7

Chapter 7
Concluding Remarks

In recent years, the development of green building is facing the transformation from "big quantity" to "high quality". The acquisition of actual performance data (including energy, IEQ and occupant satisfaction) of green buildings and its in-depth quantitative research are of great significance to comprehensively improve the quality of green buildings. However, there is a lack of high-quality, long-term and large-scale basic data on the operating performance of green buildings, which results in the shallow and incomplete cognition of performance characteristics and the failure to guide building operations. In view of the above problems, this thesis carried out data-driven research on characteristics of the operating performance of green office buildings. The main contents are as follows:

Firstly, an intelligent IEQ monitoring tool and system (IBEM) was developed to overcome the low efficiency and high cost of IEQ data collection on a large scale. The IBEM can continuously monitor air temperature, relative humidity, CO_2, $PM_{2.5}$ and illuminance with a high spatial–temporal resolution. Furthermore, it provides the graphical user interfaces for data visualization and interactive feedback. Based on the IBEM and the existing energy metering system, a long-period, high-frequency, high-density and high-precision monitoring of operating performance was carried out on 63 green office buildings nationwide. As a result, the largest green building operating performance database in China has been built for further analysis.

Secondly, the characteristics of IEQ parameters of Chinese green office buildings were revealed based on the measured data. The results show that the compliance rate of indoor thermal environment in winter is 10–40% lower than that in summer; CO_2 concentration is generally lower than 1000 ppm; $PM_{2.5}$ concentration is relatively high in Cold and HSCW zones in winter and their noncompliance rates are over 40%; Illuminance is averagely higher than 300 lx. Compared with foreign countries, Chinese green office buildings show an indoor thermal environment more in accordance with the concept of energy-saving and comfort, a lower CO_2 concentration, and a higher $PM_{2.5}$ concentration.

Y. Geng, *Data-Driven Research on Characteristics of Operating Performance of Green Office Buildings*, Springer Theses, https://doi.org/10.1007/978-981-95-2329-0_7

Thirdly, the correlations between energy consumption, indoor environmental quality and occupant satisfaction were quantitatively explored. It is found that energy conservation has both positive and negative effects on indoor thermal environment. Among them, Type-A buildings have little potential for energy conservation, and the negative effect of energy conservation is bigger than the positive effect. However, Type-B buildings still have a large space for energy conservation, and the positive and negative effects of energy conservation are almost in the same level. Therefore, Type-B buildings should be taken as the primary mining object for overall performance improvement.

Fourthly, the thesis proposed solutions to diagnose low-quality performance data based on the regression model between building monthly energy consumption and outdoor air temperature, as well as the indoor environment Kano regression model. The former can realize the diagnosis of key information such as the lighting and plug energy consumption, the cooling or heating load, the energy efficiency of HVAC system and the operating strategy, etc. The latter can quantitatively identify the influence of various environmental factors on the overall occupant satisfaction.

Finally, aiming at the problem that the massive performance data of green office buildings are hard to be analyzed by traditional methods, a research framework combining data mining and domain knowledge was established, and specific solutions were given according to the typical cases. It is found that the proposed data mining method has a good computational accuracy and interpretability, and can play an important role in pattern extraction, diagnosis, and prediction of the building performance.

To sum up, based on the data-driven idea, this study built up a high-quality database of operating performance of green office buildings, and realized the scientific and comprehensive understanding of operational performance characteristics of green office buildings. The proposed diagnostic models and data mining methods can provide technical supports for the improvement of operating performance of green office buildings.

Appendix A
The Calibration by the National Institute of Metrology

The air temperature calibration conforms to the standards of *NIM-ZY-RG-JC-122 Calibration Specification of Digital Thermometer* and *JJF 1366-2012 Calibration Specification of Temperature Data Acquisition Instruments.* The co-locating reference instrument is a first-class platinum resistance thermometer with the measuring range from – 80 to 300 °C. The calibration was conducted at eight nominal temperature conditions (8, 12, 16, 20, 24, 28, 32, and 36 °C). Measurements were averaged over 5 min.

The relative humidity calibration conforms to the standards of *JJF 1076-2001 Calibration Specification of Humidity Sensors.* The reference instrument for calibration is the M4/1311DR dew point meter. The SRH-3MC135ADR divided flow humidity generator was used to maintain a stable humidity environment. Four relative humidity conditions (20, 40, 60, and 80%) were set for calibration. Measurements were averaged over 5 min.

The CO_2 concentration calibration conforms to the standards of *GB/T 18,204.2-2014 Examination methods for public places—Part 2: Chemical pollutants.* The reference instrument calibrated the IBEM sensors in a sealed chamber (3 m^3) with an intake port, mixing fans, and an exhaust port. The co-locating reference instrument is I006 infrared gas analyzer. During the calibration, the CO_2 concentration in the chamber was adjusted, and nine conditions (481, 515, 664, 740, 881, 962, 1084, 1164, and 1264 ppm) were measured. Concentrations were held at fixed values for a period before adjusting to a new level. The averaging time for measurements is 10 min.

The $PM_{2.5}$ concentration calibration conforms to the standards of *GSH/J2011-1 Technical Specification of the Test Method for the Air Cleaner to Remove Particulate Matter PM2.5.* The calibration chamber is the same as above. Cigarettes were used to generate $PM_{2.5}$, and TSI 8533 was used as the co-locating reference instrument. Ten conditions (19, 56, 105, 162, 239, 293, 396, 453, 522, and 665 μg/m^3) were measured. The averaging time for measurements is 10 min.

Y. Geng, *Data-Driven Research on Characteristics of Operating Performance of Green Office Buildings*, Springer Theses, https://doi.org/10.1007/978-981-95-2329-0

The illuminance calibration conforms to the standards of *JJG 245-2005 Verification Regulation of Illuminance Meter*. During the calibration, the illuminance was strictly controlled by a first-level luminous intensity standard lamp. Nine conditions (30, 50, 100, 300, 500, 1000, 1500, 2000, and 3000 lx) were measured. Illuminance was averaged over 1 min for four values.

Appendix B
The Monthly Average Values of Outdoor Environmental Parameters in Different Climate Zones

Table B1 Monthly average values of indoor and outdoor environmental parameters in Cold zone

	Indoor				Outdoor			
	T (°C)	RH (%)	CO_2 (ppm)	$PM_{2.5}$ (μg/m^3)	T (°C)	RH (%)	CO_2 (ppm)	PM2.5 (μg/m^3)
Jan	26.3	13.6	576.1	43.4	− 1.7	50.0	400.0	116.0
Feb	26.5	10.2	564.6	41.1	2.3	37.0	400.0	70.6
Mar	26.9	13.5	541.3	31.7	9.0	38.0	400.0	63.2
Apr	27.9	18.0	511.1	21.7	17.5	36.0	400.0	53.2
May	29.7	26.3	494.4	21.5	23.3	41.0	400.0	48.5
Jun	27.3	40.6	528.0	25.2	25.6	50.0	400.0	42.0
Jul	27.7	55.5	566.7	36.9	27.9	70.0	400.0	52.0
Aug	27.3	54.9	555.5	31.5	26.2	71.0	400.0	38.0
Sep	27.5	43.3	513.5	31.1	23.0	58.0	400.0	54.6
Oct	26.6	38.1	494.2	51.3	12.8	70.0	400.0	84.6
Nov	25.6	22.8	529.3	49.5	4.1	42.0	400.0	99.5
Dec	25.3	16.0	553.5	43.9	− 0.2	37.0	400.0	132.7

Y. Geng, *Data-Driven Research on Characteristics of Operating Performance of Green Office Buildings*, Springer Theses, https://doi.org/10.1007/978-981-95-2329-0

Table B2 Monthly average values of indoor and outdoor environmental parameters in HSCW zone

	Indoor				Outdoor			
	T (°C)	RH (%)	CO_2 (ppm)	$PM_{2.5}$ (μg/m³)	T (°C)	RH (%)	CO_2 (ppm)	PM2.5 (μg/m³)
Jan	20.8	36.7	549.8	35.5	6.8	73.0	400.0	58.0
Feb	20.2	34.0	498.2	32.9	6.8	66.0	400.0	44.0
Mar	22.1	48.4	466.8	28.7	10.4	67.0	400.0	40.0
Apr	24.0	50.1	429.6	27.3	17.1	64.0	400.0	42.0
May	26.0	56.0	428.0	28.5	21.9	66.0	400.0	44.0
Jun	26.7	63.0	439.1	27.9	24.0	77.0	400.0	38.0
Jul	27.3	61.0	496.8	22.8	31.9	66.0	400.0	33.0
Aug	27.0	60.8	449.4	23.2	29.4	76.0	400.0	31.0
Sep	26.5	61.2	409.8	21.3	24.6	78.0	400.0	26.0
Oct	23.6	60.4	485.5	21.5	19.3	75.0	400.0	24.0
Nov	22.3	53.9	528.4	35.4	13.6	74.0	400.0	41.0
Dec	21.3	36.2	536.8	40.4	7.1	65.0	400.0	54.0

Table B3 Monthly average values of indoor and outdoor environmental parameters in HSWW zone

	Indoor				Outdoor			
	T (°C)	RH (%)	CO_2 (ppm)	$PM_{2.5}$ (μg/m³)	T (°C)	RH (%)	CO_2 (ppm)	PM2.5 (μg/m³)
Jan	23.1	45.0	540.3	24.5	15.4	81.0	400.0	62.0
Feb	20.9	39.4	467.0	22.0	15.0	75.0	400.0	44.0
Mar	24.9	49.0	464.1	18.1	17.9	84.0	400.0	43.0
Apr	25.8	51.4	487.5	17.7	21.6	83.0	400.0	40.0
May	26.1	51.0	586.4	14.0	25.1	84.0	400.0	24.0
Jun	25.9	52.4	577.8	10.9	28.2	85.0	400.0	18.0
Jul	25.9	51.2	618.7	8.5	27.9	85.0	400.0	21.0
Aug	26.5	55.8	680.4	13.9	28.7	82.0	400.0	23.0
Sep	26.3	59.0	573.0	15.1	27.9	85.0	400.0	34.0
Oct	26.2	51.9	546.7	17.3	23.6	75.0	400.0	32.0
Nov	25.2	51.4	543.8	20.3	19.3	78.0	400.0	39.0
Dec	23.5	39.3	536.0	25.4	14.4	66.0	400.0	51.0

Appendix C
The Detailed Information of the Simulated Reference Model

Table C1 The input parameters of reference model for benchmarking

Category	Content	Data source
Weather data	Outdoor air temperature, relative humidity, solar radiation, etc.	Weather station
Building shape	Building layout and partitions	Design drawings and documents
Envelop performance	Window-Wall ratio, thermal characteristics, air tightness, etc.	Design drawings and documents
Type of HVAC system	Heating and cooling source, transmission system and air conditioning terminal	Design drawings and documents combined with field survey
Indoor thermal condition	Set point of indoor air temperature and relative humidity	Reference mode[a]
Density of indoor heat source	Density of occupant, lighting, and equipment, as well as fresh air rate	Reference mode[a]
Schedule of indoor heat source	Schedule of occupant, lighting, equipment, and fresh air unit	Reference mode[a]
Operation of HVAC system	Operation timetable of HVAC system in the whole year and in a single day	Reference mode[a]

[a]Reference: Qi Zhang, Research of Operation and Usage Energy Model and Energy Simulation Method in Office Building, Mater Thesis, Tsinghua University, 2014 (in Chinese)

Y. Geng, *Data-Driven Research on Characteristics of Operating Performance of Green Office Buildings*, Springer Theses, https://doi.org/10.1007/978-981-95-2329-0

MIX
Papier aus verantwortungsvollen Quellen
Paper from responsible sources
FSC® C105338

If you have any concerns about our products,
you can contact us on
ProductSafety@springernature.com

In case Publisher is established outside the EU,
the EU authorized representative is:
Springer Nature Customer Service Center GmbH
Europaplatz 3, 69115 Heidelberg, Germany

Printed by Libri Plureos GmbH
in Hamburg, Germany